NOBEL DREAMS

NOBEL DREAMS

POWER, DECEIT, AND THE ULTIMATE EXPERIMENT

GARY TAUBES

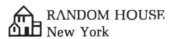

RANDOM HOUSE
New York

Library of Congress Cataloging-in-Publication Data

Taubes, Gary.
  Nobel dreams.

  Includes index.
  1. Particles (Nuclear physics)—Research.   2. Grand
unified theories (Nuclear physics)   3. Superconducting
Super Collider.   4. Rubbia, Carlo, 1934– .
I. Title.
QC7934.T38   1986   530.1'42   86-10106
ISBN 0-394-54503-6

Manufactured in the United States of America
98765432 24689753 23456789
First Edition

Book design by Charlotte Staub

To the spirit of Uncle Alfred

"Before I came here I was confused about this subject. Having listened to your lecture, I am still confused, but on a higher level."

—Enrico Fermi, 1938 Nobel Laureate in physics

"The difficult trick in the art and craft of science is to exercise discipline while still obeying one's daimon."

—Robert K. Merton, philosopher of science

PREFACE

Theoretically, Carlo Rubbia and his colleagues are engaged in that time-honored occupation, the pursuit of pure knowledge. Their work is probably more closely related to that of philosophers or, in some obvious ways, theologians than anything else. Unlike that of their brethren in solid state or nuclear physics, the work of the high-energy physicist has no practical uses. It does, however, produce spin-offs. When, in the 1860s, James Clerk Maxwell proposed that electricity and magnetism were two aspects of the same force and propagated through space in waves, what resulted was more than just that bane of freshman physics majors known as Maxwell's equations. Among the future spin-offs would be numbered electric lights and radios and television sets. Later, the spin-offs from this physics were to include such items as atom bombs and X-ray machines and computer technology.

Nowadays, high-energy physicists work in a domain of energy that is so far removed from natural earthly phenomena that it is unlikely to lead to direct technological innovations for the next few hundred years at least. The tools that they develop along the way to achieve those energies, such as superconducting technology, may have immediate applications, however. And the skills that they must learn frequently end up being put to very productive use in what is euphemistically known as defense technology; the brilliant minds of the Manhattan Project were physicists by trade.

Whether or not this all turns out to be infinitely regrettable, the high-energy physicist no more plies his trade to enhance the technological level of his nation than did Ahab set sail in the *Pequod* to

enrich the coffers of the Nantucket spermaceti industry. The immediate end product of the work of the high-energy physicist is knowledge, pure and simple; or, to be more precise, the answer to a single,
ultimate question.

CONTENTS

INTRODUCTION

On December 10, 1984, Carlo Rubbia finally got his Nobel Prize.

The award ceremony begins at four in the afternoon. The city has been pitch dark since the sun set two hours earlier. The blue-gray concert hall is dimly lit; the cobblestone plaza in front slowly fills with enormous black limousines. Inside, the audience is in formal attire. On a blue carpet at the center of the stage, a light blue circle rings a light blue "N"; on a dark wooden dais is affixed a golden relief of Alfred Nobel's bearded profile. At 4:15 P.M., the orchestra begins tuning up. A French horn runs through the strains of "Morning Mood" from *Peer Gynt* as the audience files in. The king and queen of Sweden enter at four-thirty and the crowd rises in their honor. Then the laureates march in, Rubbia first. They take their places in front of red-padded armchairs, and when the king sits, they sit.

Rubbia settles with his hands between his legs, twiddling his fingers, looking out over the crowd. He is fifty years old. He is a large man, and over the past five years he has gained more weight than his frame can comfortably support. He has sandy brown hair that falls in a lock over his forehead, and a sallow, puffy complexion that, more than anything, is the result of weekly intercontinental commutes between his laboratory in Geneva and his office at Harvard. He has light blue eyes that are his most striking feature. They are almost inhumanly bright.

Carlo Rubbia is arguably the most powerful man in high-energy physics. He is an experimental physicist. His is a discipline in which political savvy, physical endurance, money, and maybe guts, can be as important as scientific insight. In that arena, he might just be the best there is. But he is renowned for his frenetic energy and his inability to sit still—or even to stay in one city or one country for more than a week—as much as for his physics and his political acumen.

On stage, he is as jittery as can be, sitting in full view of the king and queen, seventeen hundred guests, and a national television audience. He clenches and unclenches his hands, crosses and uncrosses his feet. The other laureates are more relaxed—the British economist even looks bored with it all. The dignitaries seated in a gallery at the rear of the stage beat their black patent leather shoes softly to the *Karelia Suite* of Sibelius.

The chairman of the board of the Nobel Foundation opens the ceremony with an address on the theme of international cooperation. He is followed by a physicist of the Royal Swedish Academy of Sciences, who delivers the presentation speech for the prize in physics. He says that Rubbia and his Dutch colleague Simon van der Meer have been awarded the 1984 Nobel Prize for, in the official wording, "their decisive contributions to the large project, which led to the discovery of the field particles W and Z, communicators of weak interaction." The two scientists stand up. Rubbia tugs his white vest down and straightens his jacket. "The discovery of the W and the Z is not the end," reads the speaker. "It is the beginning." And then to Rubbia, "I now invite you to receive your prize from the hands of the king."

Rubbia walks slowly forward to take the gold medallion that he has pursued relentlessly for many years. He shakes hands with the king and bows to him. He bows to the dignitaries and to the audience again and again, then returns to sit at the foot of the stage. He fidgets a little impatiently as he holds his Nobel Prize and waits for the other laureates to receive theirs. He has finally reached his destination.

I first met Carlo Rubbia in the winter of 1982. I had gone to Harvard to interview Sheldon Glashow—a part owner of the 1979 Nobel Prize in physics—for a magazine article on the new generation of atom smashers. At the time I was a neophyte science writer, and I believed that I had to quote several Nobel Laureates per story in order to establish my credibility. This was also an unearned opportunity to hobnob with the wizards of the upper stratosphere. I had studied physics at Harvard, but had never mingled with anyone above the rank of assistant professor. Glashow politely answered my questions and then just as politely suggested that I was talking to the wrong man. I should speak with the guy down the hall, he said, Carlo Rubbia, because he was a shoo-in for the Nobel Prize in two years.

I knocked on Rubbia's door and he invited me in. I sat down, pulled out pen and notebook, and for the next hour scribbled maniacally as Rubbia paced and ran his hands through his hair and mixed one metaphor after another while generally holding forth on his version of the gospel of experimental physics. He reminded me of a fundamentalist preacher, but one who appeared to believe that if the Word of the Creator had been heard only in the first fiery instant of creation—and had since vanished from the universe—then all he needed was enough money and enough genius and he could build a machine that would bring it back.

I could not discern the slightest trace of modesty in Rubbia's ambition. His work, as he described it, was simple: to pick the universe apart, find out of what it was made, and why it was what it was. This was unfortunately complicated by the fact that the universe is not now made of the same stuff as when it was created 15 billion years ago, but rather of the descendants of that stuff. If the universe had indeed begun in the unimaginable fireball of the big bang, then it had been at its very hottest at the very instant the bang began, and had been cooling down ever since. The result was that whatever had happened and existed in that moment of creation had since been unalterably changed in the 15 billion years of slow cooling that have followed. Physicists are rather like curious diners faced with a cosmic stew and no idea of what the original ingredients were.

Accelerators—atom smashers, in the old lingo—are like time machines with which high-energy physicists can poke beams of energy back into the primordial big bang. The more powerful the accelerator, the farther back they can get. Rubbia and his colleagues at the European Laboratory for Particle Physics in Geneva, known as CERN, have built a machine four miles in circumference that smashes two beams of particles head on, ramming a nearly infinite amount of energy into an infinitesimal point in space—a point that then becomes for nearly no time at all as hot and as violent as the universe was one trillionth of a second into the big bang. Whatever strange breeds of elementary particles that once existed in that initial conflagration can now be created out of the energy released in Rubbia's extraordinary collisions. ("One of the things we do," he told me, while soaring to one of his more humble metaphors, "is produce new matter. The man on the street thinks all matter is made by God in the moment of creation. High-energy physicists like us are repeating over and over again the miracle of creation.") And Rubbia's colliding-beam machine is ten times more powerful than any other accelerator in the world.

Rubbia and his colleagues went on to build a second gargantuan piece of machinery: the experiment itself. This is $20 million worth of iron and electronics, weighing 2,000 tons, that peers into the inferno as though it were the most powerful microscope in the world, hooked up to the highest speed camera. If the accelerators are the time machines, the experiments are the robot eyes and brains that allow the experimentors to see and understand what it is that their accelerators have made.

Specifically, Rubbia was looking for the signs of two particular elementary particles called intermediate vector bosons—"field particles W and Z, communicators of weak interaction," as the Nobel Foundation would put it—that were so heavy, in terms of high-energy physics, that hitherto they could not be created in an accelerator. (In Einstein's universe, energy and mass are equivalent. The more massive a particle, the more energy is needed to create and identify it.) "By the end of the year," Rubbia told me, "we will know the Truth."

And by the end of the year, they did.

Physicists have always been enamored of the ideal of unification, and have pursued it, as Tennyson might have said, like a sinking star. In so doing, they have concocted a recipe for the universe that consists of a bare minimum of ingredients. They acknowledge the existence of only four forces, which, in one combination or another, can account for any and all actions witnessed in the universe. For the universe at large, there is gravity. For the microscopic universe, there is electromagnetism: the strong force, which holds protons and neutrons together in the nuclei of atoms, and the weak force, which is responsible for certain forms of nuclear radioactivity. Along with these forces, physicists have identified seventeen different species of particles, with which they could account for every one of the hundreds of particles either the universe or their atom smashers have ever thrown at them: five particles that transmit forces (photons for electromagnetism, gluons for the strong force, and two W's and a Z for the weak force); and twelve particles that make up all matter (six breeds of quark, the lightest two of which make up the protons and neutrons; and six known as leptons, which include three types of electrons and three massless, charge-less particles known as neutrinos).

Physicists have constructed piece by piece the theories that define the actions and existence of these fundamental particles and forces of the universe in terms of consistent mathematical equations. And their ultimate desire is to have one single theory—a theory of everything—that would account for the actions of all four forces, and explain the existence of all seventeen species of particles. By the 1960s, they had made extraordinary progress. They had a theory for gravity, Einstein's relativity, which described the actions of the universe at large scales. They had one working quantum theory, called quantum electrodynamics, or QED, that described the workings of electromagnetism. And they were plugging away at theories for the weak and strong forces.

Throughout the sixties, one scattered coterie of theorists worked on a quantum theory for the strong force, which would come to be known as quantum chromodynamics, or QCD. Simultaneously, another crew was laying the cornerstones for a quantum theory of the weak force that, if it could be experimentally proven, would reconcile the actions of the weak force and electromagnetism in the same way that, a century earlier, Maxwell had decreed that electricity and magnetism were different aspects of one and the same force. Together, these would become known as the "standard model."

The key players in the construction of the electroweak theory—or at least as the Nobel committee would see it in 1979—were Sheldon Glashow, who, back in 1961, had published the theoretical framework in an obscure journal; Steven Weinberg, then at MIT; and Abdus Salam, then of Imperial College in London, who had pulled the critical pieces together in 1967 and 1968.

The Weinberg–Salam model, as it was known, was ignored until 1971. It had the flaw, endemic to many failed quantum theories, that it could not be used for calculating the actions or interactions of particles and the forces in question—which, unfortunately, was exactly what physicists would want to do with such a theory—without resulting in meaningless and infinite answers. This failure was known as non-renormalizability. But in 1971, a brilliant Dutch physicist named Gerard 't Hooft managed to renormalize the theory—or un-non-renormalize it, as the case may be—for *his* Ph.D. thesis, and suddenly the world had a working theory of the weak force. Weinberg's 1967 paper, which had been cited zero times that year, was cited a couple hundred times in 1972.

One of the predictions of this theory was that there existed these two charged-force-carrying particles, the W's—which had been part of all attempts to make a theory of the weak force since the 1930s —and a neutral one, the Z-zero, which was new. These would transmit the weak force in the same way that photons carry the force of electromagnetism. (Imagine infinitesimal bundles of energy that are tossed back and forth, like medicine balls, between particles.)

The theory predicted that the W and the Z would be the heaviest of all the particles, and hence the most difficult to create in a particle accelerator. Rubbia came along in 1976 pitching a fanciful idea of how to create them by revamping an existing accelerator so that it would collide protons and their antimatter counterparts, antiprotons. He pushed his idea first in America, and then at CERN, which, run by a consortium of thirteen nations, is the largest European laboratory. The Europeans built the machine, and it worked in what seemed like no time at all. In January 1983, Rubbia announced that they had discovered the W particles. And in June he announced that they had found their Z's.

The ramifications of Rubbia's success spread far beyond the halls of academia and the pursuers of pure science. Rubbia became ensconced as the most powerful figure in high-energy physics, and this scared the hell out of the U.S. physics community, as well as the bureaucrats in Washington.

Until Rubbia came along, the United States had owned high-energy physics. From 1950 through the end of the last decade, Americans had made virtually every major discovery and won nearly every Nobel Prize. Of course, many of those physicists were European immigrants, but as far as the United States was concerned, they were now Americans. No matter how tough the European competition was in physics, the Americans always beat them. In the seventies, European physicists seemed to be regrouping. Yet even though CERN had the most powerful accelerator in the world by a factor of ten, and a budget as great as all of the U.S. labs put together, the American physicists still somehow came up with the three most notable and clearcut discoveries of the decade, and all the Nobel prizes.

When Rubbia's machines produced the W's and Z's in 1983, it marked the waning of American domination. Physicists had been looking for the bosons for decades—they liked to call them their Holy Grail—and now they were found in Europe by European physicists. Five days after the announcement of the discovery of the Z (also known as the Z-zero because it has no electric charge), *The New York Times* ran an editorial so virulent that many American physicists found it embarrassing. "Europe 3, U.S. Not Even Z-Zero" was the headline. It demanded revenge, and not just any revenge, but "earnest revenge." The newspaper of record in America was demanding revenge against an experiment that had been done in the spirit of the pursuit of pure knowledge.

Although European physics had been rebuilding for two decades, it sometimes seemed that Rubbia had managed almost singlehandedly to switch the power in that field from the New World back to the Old. One week after the *Times* editorial, an advisory panel of physicists to the Department of Energy voted to cancel the work on one technologically troubled accelerator then under construction (which was like poker players folding a hand after throwing a hundred million dollars into the pot) and to go ahead full speed with the design and construction of a machine that would be a hundred times more powerful than the one CERN had built, and at least twice as powerful as anything the Europeans could conceivably build in the future. It would cost $5 billion, at least. The President's science adviser, George Keyworth, told Congress, "I won't conceal my opinion that it would be a serious blow to U.S. scientific leadership if that facility were built in another country." It was intended to be the last word in accelerators. They called it the Superconducting Super Collider—the SSC.

Meanwhile, Rubbia's amazing colliding-beam machine was still running. In April, 1984, at a meeting of the American Physical Society in Washington, Rubbia told reporters and physicists that he was on to what might be the biggest breakthrough in physics in thirty years. Something that would make the Nobel Prize–caliber W and Z pale in comparison. It might be another particle; it might be another quark; it might be almost anything. He had already seen the first signs, and he was waiting for his machine to provide confirmation. And no other machine could do it.

"It is really true," Glashow had said, "that the basic questions of physics come only from one place. We have all become Roman Catholics of a sort; our pope is Carlo. And of course anything he says is incontrovertible because nobody else has a device which can controvert him."

The fact that Rubbia, with all his political acumen, was not the man that many of his colleagues would like to see in a position of power seemed to bother the high-energy physics community not at all. He got things done. He achieved.

Rubbia is undeniably brilliant. No one who meets him fails to be impressed with the quickness of his mind. James Bjorken, one of the premier theorists in the United States, and a very smart man, put it this way: "He's just one brilliant guy. Whatever else one thinks about him, he's really smart." But what about whatever else one thinks about him? As I talked to high-energy physicists, I found that the closer I got to Rubbia himself, the more vitriolic became the feelings about his personality. As with most public figures, his public and private images had little in common. At the Department of Energy, the branch of the government that provides the cash for most high-energy physics endeavors, one official told me, "I would say that with very rare exceptions, Carlo is a most congenial and pleasant person." This is witness, if nothing else, to Rubbia's talent as a politician, since from his peers I would hear virtually the same line repeated over and over: "Rubbia is brilliant, but I wouldn't want to work for him."

And then there were those who did work for him. From his co-workers, I heard version three. When I tracked down some of the students, post-docs, and assistant professors Rubbia had taken on at Harvard, I found that many of them had before long left Rubbia, or left physics altogether. One dropped out and went to business school. Another dropped out and floated around the California drug crowd; a third drove a cab in Cambridge for years; several just disappeared. I found one working successfully in industry in California, who told me that he had been in love with physics until he met Rubbia, and that anything bad I had heard about the man was probably true. Of those who are left in high-energy physics, few still have pleasant dealings with Rubbia. One told me, "You need a skin like a lion and a heart like Jesus to work with him." Another said, "He's just a crazy man." He added, however, that while "the field of physics would clearly be a disaster if more than a handful of people were like him, it just as clearly would not have gotten this far if not for a handful of crazy men just like him."

I heard the same story in Europe. Of two of the physicists who had worked most closely with Rubbia on his *pièce de résistance* at CERN—two of the most talented young physicists in Europe—one of them, Bernard Sadoulet, told me that he compared Rubbia to a black hole "warping the universe around him," and that "obviously Rubbia has read and assimilated Machiavelli." The other, Mario

Calvetti, simply said, "Even after Rubbia won the Nobel Prize, he took pleasure in hurting people. From the human point of view, he's a disaster, a complete disaster." Both of them have quit the experiment. Of the 150-plus physicists I talked to, only a handful would say nothing at all either critical of Rubbia's modus operandi or derogatory about the man himself. Several of them told me that if Rubbia ever heard what they had told me, they would not work in physics again in Europe until after Rubbia retired.

One assistant professor, Larry Sulak, had run afoul of Rubbia in the late seventies when he tried to compete with the Italian for funds on a "hot" experiment. Rubbia had, among other things, written a letter of dis-recommendation to the university Sulak turned to for tenure. When I asked Sulak why he still would work with Rubbia after that, which at times he claimed he would, he told me in his West Virginia accent, "Carlo's a brilliant physicist, no matter what kind of an asshole he might happen to be."

As a physicist, Rubbia is an electronics wizard. He has an eye for which experiment to do, and how to do it. He has a fondness for playing with the newest technology, and a mind that can throw out ideas like a Roman candle. And he pushes them all with equal enthusiasm until he loses interest or his colleagues kill the ideas that are obviously chimerical. And whenever a laboratory builds a new accelerator with a higher energy than ever before, Rubbia always seems to be able to get access to the highest energy beams first. This is all unarguable.

Rubbia has a passion and a curiosity that burns with such fire that he himself describes it as inextinguishable. "For the moment," he has said, "my desire to get somewhere is so large, so strong, that I cannot conceive of seeing myself turned off." He is obsessed. Rubbia's schedule, if nothing else, prohibits his engaging in anything but physics. He takes vacations every couple of years or so, but usually complains about feeling bored and claustrophobic. He has rarely been known to take a single day off: not Sundays, not holidays. Maybe a few times for special family occasions, but none of his colleagues can point them out. He sees his wife and two children maybe half a dozen hours a week; when asked what his wife thinks of this, Rubbia once commented, "I was hopeless already when I was young, so she knew what she was getting into. Perhaps she might have expected some slight changes over the years, which haven't occurred." He reads Agatha Christie mysteries when he flies, because he likes the idea of Poirot facing a seemingly insoluble mystery and solving it. The universe is a lot like that.

Rubbia has not always been at the top of his profession. In fact, in the seventies he was as famous for his mistakes as for his successes. ("When he is told about something good, he finds it," Glashow said once. "And when he's not told, sometimes he finds things that aren't really there.") And whenever Rubbia's experimental results were proven to be wrong, he would shrug off the criticisms, as though being wrong in a field where the whole aim was to be right didn't faze him in the least. It was not until the discovery of the W and the Z that Rubbia finally exonerated himself. And after all, as Larry Sulak says, personality is not physics. Or as Georges Charpak, one of the most renowned of European physicists, put it more cynically: "Carlo is a high-energy physics animal, perfectly adapted to the milieu. Those who complain about this are no longer adapted to the milieu."

By 1984 the physicists of Europe, perhaps subscribing to this Darwinian philosophy, were talking about making Rubbia director general of CERN. They wanted an accelerator that could compete with the American SSC, and they thought that Rubbia could build it for them. ("He has the fame," Ugo Amaldi, a senior CERN physicist, told me. "He has the power. On the human side, he has a difficult character; he makes people suffer, some of them very much. But on the other side, to have a Nobel Prize winner as the DG of CERN is a very good asset for the organization.") They had heard his gospel and they believed.

While these deliberations are being made, Rubbia is walking around Stockholm with the air of a man who has more surprises up his sleeve. He gives the distinct impression that his Nobel Prize is just a step along the way, a necessary part of some grand scheme. He scoffs at suggestions that the so-called Nobel Effect will set in —that he will spend the remainder of his life lecturing on the evils of modern times and writing lengthy correspondence to fellow great men. Instead, he hints that he has already in hand his next major breakthrough, and that nothing will slow him down. His accelerator has produced a great many things since they wired it up and threw the switch in 1981. It has produced a Nobel Prize, and as far as Rubbia is concerned, it was likely on its way to producing another.

At his Nobel lecture, which Rubbia gives on a Saturday morning at the Swedish Royal Academy of Science, he describes the accomplishment that has won him this prize. The lecture is in a modern white auditorium, full of students, professors, and intrigued amateurs. Rubbia wears a check sports coat and his hair falls over his forehead. He begins as he often does when lecturing, in a sort of herky-jerky way, like an old car on a cold day, shooting forward in little spurts and then catching and coughing before shooting forward again.

The expressed purpose of the lecture is to describe the work for which he has been awarded the prize. He describes the huge atom smasher that straddles the border between Switzerland and France. He describes the 2,000-ton, house-sized piece of electronics that allows his team to see the head-on collisions between particles moving at nearly the speed of light. And he describes how they have seen the unmistakable signs of the field particles W and Z, communicators of weak interaction.

As Rubbia goes along, he picks up speed until he is talking at upwards of two hundred words a minute, his usual pace, and thumbing a new transparency onto the overhead projector every fifteen or twenty seconds. Then, about twenty-five minutes into the lecture, he digresses; whether it is important is impossible to tell. Certainly few in the audience catch on.

Rubbia has been talking about his W particles and referring to the few that have been found, in the jargon of high energy physics, as "events." "However," he says now, "in addition to these events, we also find others. Those are different than the W particles, somehow a new physical phenomenon. I have no time, nor is this the place, to discuss it, but I have to point out the presence of such phenomena in order to understand better what we're doing." Then he goes on to say, "The nature and the origin of these events are unknown. For the moment they are discarded, and we continue our story of the W particles."

That is it for the day. The nature and the origin of these events are unknown. What does it mean?

Three days later, the two physicists of the Nobel class of '84, Rubbia and Simon van der Meer, are joined by the three from the class of '79—Weinberg of the University of Texas, Glashow of Harvard, and Salam of the International Centre for Theoretical Physics in Trieste —for a roundtable discussion on the status of high-energy physics. It is at the roundtable that Rubbia's bizarre events that were not W's or Z's, whose nature and origin are still unknown, keep slipping into the conversation.

The laureates talk about the future of physics. They hit every angle, theory and experiment, past and future, and they keep sliding back to Rubbia's mysterious events. They discuss the current theories of elementary-particle physics, which can explain everything seen in physics from radio waves to laser beams, everything from a trillionth of a trillionth of a second after the big bang to the present and perhaps through the very end of the universe. They can explain almost anything, they said, except maybe Rubbia's mysterious events.

They discuss the problems with creating an even more powerful theory, a more fundamental theory, a theory that would be able to explain the state of the universe even before that trillionth of a trillionth of a second after the big bang, during the hottest, most extreme moments of its birth; maybe even before its birth, if there was such a before. A theory of everything, they call it. Weinberg describes this search for an ultimate theory as almost entirely frustrating. Salam says it is a sad situation. And Glashow remarks that no one is getting anywhere. But they keep coming back to Rubbia's strange events, and it is Glashow who finally says simply that Rubbia has been very modest, but that he is "sitting on a bomb."

It is Glashow who, when asked to whom he thinks the next Nobel Prize in high-energy physics will go, replies that it will go to Rubbia "for the discovery of something that he has not yet begun to dream of."

BOOK I

RENEGADES, MADMEN, AND THE END OF THE ALPHABET

"Once down is no battle."

—Jack McAuliffe, the last
  bare-knuckle lightweight
  champion of the world

1.

"A.C.D.C."

"One theorist told me we set back physics for twenty years, and I said, 'Hell, you're exaggerating. We only set back physics ten years.'"
— Don Perkins, an Oxford physicist, on one of the numerous erroneous results of experimental physics

Carlo Rubbia once told me that his preferred definition of an expert was that of the man who has made all his mistakes already. I had no doubt that, as far as he was concerned, he was referring to the 1960s and the first half of the 1970s. They were his dog years, or his learning years, depending on how you looked at them. There was a revolution going on in physics, and although Rubbia always managed to be at the front, it was as if all the big victories eluded him. Or he fought the wrong battles with the wrong armies. He won no medals, and came out of the fighting with his reputation battered.

Through that time he became known as one of the *enfants terribles* of high-energy physics. Brilliant, but the butt of jokes. A tempestuous troublemaker. "He's a very human guy," Don Perkins once told me. Perkins had worked and competed with Rubbia for twenty-five years. "I'd prefer Carlo one million times to somebody like [he named two other Nobel Laureates], who have no sense of humor at all. But Carlo can laugh at himself, I think. Maybe he only does it in private, but he does it. I'm sure."

That this feeling is pervasive among physicists is evident, as much as anything, from the fact that they always refer to Rubbia as Carlo. "Well, that's Carlo . . ." they'll say. What more can be said?

Rubbia was born in 1934 in Gorizia, Italy, a small town north of Trieste on the Yugoslav border that until World War I was part of the Austro-Hungarian Empire. The people were Teutonic, an overflow from the Alps. Rubbia himself looks more Slovak than anything. His father worked for the telephone company and his mother taught school.

Rubbia grew up during World War II, which rolled through his countryside, ravaged his people, destroyed his home, and provided him with a child's playground. "The war represents a tremendous tragedy for everyone except children, who don't feel it," he once told me. He spent the years before the war—or so his mother was quoted in an Italian fashion magazine—playing with Meccano sets. He spent the years after it scavenging radio equipment that had been abandoned as various armies marched through on their various advances and retreats. He took the radios apart and put them back together; he built antennas, and he jury-rigged new receivers out of the debris of old ones. He became such an electronics freak that he still remembers the day the newspapers announced the invention of the transistor as a great day in his life.

By the time Rubbia arrived in college, his skills had reached mythic proportions. He went to the Scuola Normale Superiore, a free school that takes ten out of five hundred fifty applicants every year and is run in conjunction with the University of Pisa. Enrico Fermi came out of Pisa, which is recommendation enough in Italy. The Scuola Normale is run like a monastery, the students locked inside with nothing to do but study. It was, Rubbia said, "exactly what I wanted. You have to perform, and if you don't, you're kicked out immediately." Aside from knowing more about electronics than any undergraduate ever, Rubbia gained renown for his abilities as a human voltameter. He would stick his fingers in electric sockets and tell the correct voltage. He was accurate to within 10 percent, and he never killed himself. His schoolmates still remember the voltages. They were impressed.

After his Ph.D., in 1958, Rubbia went to the United States to do his post-doc at Columbia, which was far and away the hottest spot for any young physicist. It was one of the most concentrated collections of talent in physics ever: T. D. Lee was there, who won the Nobel Prize in 1957 with C. N. Yang for disproving the law of conservation of parity; Madame Chien-Shiung Wu, who discovered what is known as parity violation; Jack Steinberger, who is now considered by many to be among the greatest experimental physicists (his former colleagues actually say things like "Jack Steinberger, the greatest"); Dick Garwin, an experimentalist who might have been better than Steinberger and quit to become, among other things, a world expert on nuclear arms; Charles Townes, who developed the laser and won the Nobel in 1964; Mel Schwartz, who invented neutrino physics and discovered the sec-

ond neutrino with Steinberger and Leon Lederman; Lederman himself, who went on to become director of Fermilab and to discover
the fifth quark; Nick Samios, who became director of Brookhaven
and discovered the omega minus particle; and Steve Weinberg,
who would put together a theory of weak interactions and share
the Nobel in 1979. This congregation was led by Isidore Rabi, who
won the Nobel in 1944 and is one of the grand old men of nuclear
physics and the Manhattan Project.

All of these intellects would meet every Friday for tea and
bounce ideas off each other. And everyone was studying "weak
interactions," which is what physicists called the fundamental force
that was responsible for certain types of nuclear radioactive decay.
"It helped develop everybody's tastes," is the way Nick Samios put
it.

Rubbia said that when he arrived at Columbia, he felt like a rube
who had been air-dropped into Times Square on New Year's Eve.
If he was aggressive before Columbia, he just turned up the heat to
match the competition. He also developed the craving for weak
interaction physics that seemed to afflict all his colleagues. He
worked with Steinberger, and after two and a half years was offered
a position as an assistant professor. But in those days, assistant
professors were buried in the lower levels of the hierarchy with
graduate students and laboratory mice.

Rubbia moved back to Europe to work at the new European
laboratory known as CERN. It had been built out of the ashes of
World War II on the basis that every country involved would be an
equal partner. The countries included, among others, Germany,
France, Italy, and Great Britain.

Through the sixties, Rubbia's reputation at CERN was equal
parts very, very good and very, very bad. He had become a group
leader, the equivalent of a tenured professor, while he was still in
his twenties. This was unheard of among European physicists. But
Rubbia had his electronics wizardry. ("Carlo thinks in a sort of
orthogonal way from most people," one physicist told me. "When
he is thinking about an experiment, he is thinking about the actual
electronic circuit that will make it work. It's all a unity to him
somehow.") He had a seemingly limitless fund of ideas about what
experiments to do, and, maybe more importantly, what experiments
other physicists had not yet done, and he knew how to make them
work. He was always aware of the latest technology, and he was
ready to risk everything to use it. CERN's director general at the

time, Victor Weisskopf, promoted Rubbia against the opposition of his staff.

Rubbia had an incurable passion for physics. If his proposals were rejected by the management, he would do experiments under the table. He would set up his equipment on test beams and, if questioned, explain that he had only been checking his apparatus. It was, as Georges Charpak put it, "a proof of love for the subject."

He had more physical energy than most mortals, and his body seemed to be able to take any abuse to which he subjected it. (In 1969, a Princeton physicist had written a paper in which he said that a certain type of physics could only be done by hyperactive physicists, and he cited Rubbia as the perfect example.) Rubbia could yell louder than any of his colleagues at CERN, and usually think faster, and the result was that he rarely lost an argument. Even when he was wrong, which was not infrequently, it was impossible to tell until the smoke had cleared, and that might be hours or weeks later. He was capable of walking out of a meeting yelling obscenities, then turning to the first person he met with a smile. As one young Welsh physicist said, "A small amount of contact with Rubbia was more than the recommended adult daily dose."

He was considered one of the three toughest men to work for at CERN, and as far as I can tell, few physicists who worked for him liked him. According to Victor Wiesskopf, maybe half of the people who worked for him eventually quit because they could stand him no longer. He was unsteadfast. Frequently he failed to finish what he had started. He tended to create extravagant experiments, then leave them for other extravagant experiments as soon as they showed signs of coming up with unextravagant results. He had a reputation for impetuosity, for lacking the patience to do the kind of excruciatingly careful analysis high-energy physics demands. And physicists who knew Rubbia at the time suggested that, as a result, his numbers were inaccurate as often as not. "I would feel badly if I did something wrong," explained one senior physicist who worked with Rubbia in the sixties. "For Carlo, that's not what counts. Clearly, he has a different kind of thing that drives him." Bernard Sadoulet, who first met Rubbia in 1969, put it more bluntly: "His numbers are what they are. They are usually wrong—but if they suit his purpose, nothing is wrong."

There is one brutal rumor to this effect that dates back to 1964. A California physicist told me he heard that Rubbia's group had performed an experiment to measure the parity of a subatomic particle known as the psi hyperon, and that when the experiment failed, Rubbia faked the results and reported them at a CERN seminar as real data. When I tracked the rumor back to CERN, I found two prestigious European physicists who had worked on this experiment, and both confirmed it. "We were looking for an effect that we didn't find," one of them said. "And yet Carlo gave a seminar. I don't know where he got the numbers from. I guess he must have invented them." The other told me that Rubbia had advised them they should never admit that any experiment had been a failure, because it would ruin their careers. And he went on to say that they had had to beg Rubbia not to give this bogus seminar at Berkeley. A third member of the experiment, Georges Charpak, did not deny this account, but added that he thought it would be trivial to mention it, because high-energy physics is full of stories like that. Wiesskopf, on the other hand, explained to me that it was not unusual for a group leader to be optimistic about the results of his experiment, and to present them in a seminar even if the experiment was not yet finished. The key factor in Rubbia's defense, said Weisskopf, was that he had not published these results.

However much of this was known to the physics community, it was either overshadowed by Rubbia's brilliance or ignored.

In 1970, Rubbia was offered a full professorship at Harvard. He had been recommended by an American physicist who had seen his work at CERN, and who knew that Rubbia had been rejected for a tenured position in Italy (because he had no teaching experience, according to Rubbia) and that he was on the market. Rubbia took the post.

It is probably not a coincidence that Rubbia's reputation nose-dived soon after he decided he could run experiments both at laboratories in the United States and in Geneva, teach courses at Harvard, and commute between the three. At times, he had four separate experiments going at three geographically distinct laboratories, and none was showering him with glory.

In 1971, CERN switched on the most powerful accelerator of its day, the ISR—for Intersecting Storage Ring—the first proton-proton collider ever. It was an extraordinary machine that would hit the very highest energies ever created. Rubbia wanted to be first on the machine. "We had to observe the first collisions," he said later. "The first moment those beams go around. I couldn't see myself waiting for a year." But the CERN management had given priority to a competing experiment.

Rubbia went through contortions to get precedence. He tried to have detecting equipment built quickly in the machine shops at CERN, but was informed that his competitors had priority there, too. He then borrowed some equipment from a Harvard colleague, flew it from Boston to Geneva checked as luggage, and rolled it into the path of the colliding beams while the technicians were taking a half-hour tea break. He took the first pictures of protons in collision, and showed them a few weeks later at the American Physical Society meeting in New York. It was called a tour de force.

Not too long after this, the ISR physicists engaged in their only important piece of physics of that period: a measurement of a parameter known as the proton-proton cross section which turned out to rise as the energy of the beams increased. It was a difficult experiment, because none of the physicists had worked before with colliding beams of protons. They threw together quick and dirty experiments and then they rushed the measurements. Rubbia reported erroneously that the cross section remained the same as the energy increased. His competitors reported that it rose. Rubbia argued with his competitors, insulted them, and even tried, unsuccessfully, to convince the CERN management that he was right and that his competitors should be restrained from publishing and forever blackening the name of the laboratory.

It was the first of three mistakes Rubbia made in the seventies that would leave him with a reputation best encapsulated by the American physicist Robert R. Wilson, who in a moment of major irritation called Rubbia a "jet-flying clown."

From Geneva, Rubbia redeployed his efforts to the Illinois prairie land west of Chicago, where, for political reasons, the next great atom smasher in the world was being built.* This laboratory is now called Fermilab. Its machine, known as the main ring, is four miles in circumference. It spins protons to 400 billion electron volts, or 400 GeV,† and then shoots them down beam pipes, where they crash into targets and create secondary beams of particles that are used for physics experiments, or as the physicists themselves refer to

---

*According to Robert R. Wilson, who built the lab and should know, Lyndon Johnson traded it in a classic piece of pork barrel politics to Everett Dirkson, when the Illinois senator agreed to withdraw his opposition to a project to establish Russian consulates in major U.S. cities.

†Both the energy of accelerators and the masses of elementary particles are referred to in terms of electron volts, or ev, the energy gained by an electron when accelerated between the poles of a 1-volt battery. A GeV is a billion electron volts. A TeV is a trillion.

their work, "physics." Rubbia and two American colleagues, Al Mann of the University of Pennsylvania and Dave Cline of Wisconsin, had the very first experiment on a beam of neutrinos, which was the most powerful in the world. The collaboration was known as HPW, for Harvard, Penn, Wisconsin.

Since neutrinos interact only through the weak force—no other force has any power over them—they are the perfect mechanism for studying weak interactions. Neutrinos are so ethereal that they can pass through light-years of solid matter before colliding with the core of an atom and betraying their existence. If neutrinos were not so easy to make, this characteristic would have been troublesome to physicists. But the Fermilab beam could produce billions of neutrinos every second. Anyone who wanted to do neutrino physics could see a neutrino collide with an iron atom once every couple of seconds.

If such a phenomenon as a guaranteed discovery existed in high-energy physics, it was in 1973 on the firing line of that Fermilab neutrino beam. It was not just that it was the highest energy neutrino beam in the world, operating in an energy range that had never before been studied; it was that by 1973, the experimentalists finally had a viable theory of the weak force with which to work, the Weinberg–Salam model, and that theory could describe exactly what phenomena would be created. Specifically, the theory predicted that the neutral Z boson should be eminently detectable in neutrino collisions with matter through something they called neutral currents.

The existence of neutral currents would be indirect proof, if found, that the Z existed. It would be direct proof that the theory was correct, or equally direct proof that it wasn't. It was also the kind of profound, definitive experiment that seems to impress the Nobel Foundation. Steven Weinberg himself called Rubbia and told him that he should go look for neutral currents, and Rubbia, who knew Weinberg from Columbia, took his advice. One year later, however, the only definitive result that had come out of this experiment was that the theorists had obtained their initial proof of the correctness of the Weinberg–Salam model through the discovery of neutral currents. And that proof did not come first from Rubbia's Fermilab collaboration. Instead, it came via a collaboration of Europeans working at CERN with an enormous but antiquated bubble-chamber experiment, known as Gargamelle.

In June of 1973, after one year of analyzing data, the Gargamelle collaboration had concluded—albeit with much internal dissension—that their data indicated the presence of neutral currents, and they were on the verge of announcing the discovery. The HPW experiment had taken data for a short period six months earlier, and the physicists had been working at the analysis since then. For most of that time, however, Rubbia had ignored the experiment, being embroiled in his own controversy at CERN over the proton-proton cross-section measurements. When Rubbia heard that the Gargamelle collaboration had neutral currents, he set Larry Sulak, who was at that time one of his Harvard post-docs, to a quick and dirty analysis, the conclusion of which was that HPW had also bagged the neutral currents.

Rubbia wrote André Lagarrique of Gargamelle, saying that the Fermilab experiment had one hundred unambiguous neutral-current events, and that "we are in the phase of final write-ups of the results." He requested that when the European collaboration announced the discovery, they might be kind enough to mention that HPW had equivalent results. The letter could be read either as a bluff or as a delaying tactic.

Lagarrique showed Rubbia's letter to Don Perkins, head of the English contingent at Gargamelle. "I burst out laughing," Perkins recalled. "He said, 'Well, what are you laughing at?' And I said, 'Well, that's Carlo. He's got hundreds of events definitely showing the existence of neutral currents. Then what the hell does he need Gargamelle for? Why doesn't he just go publish them? What's he writing to you for? He needs something. It's ridiculous.' "

Instead of debating for another few months the merits of publishing a paper, as they had appeared ready to do, the Gargamelle collaboration decided to announce their results immediately and take their chances with posterity. Rubbia's bluff had been called.

Rubbia, meanwhile, was having a bad summer all around: Back in Boston, he was asked by U.S. Immigration to leave the country after his visa had expired and he had failed to file for an extention. The immigration authorities gave him a handful of days to see to his affairs and leave the country. On his last trip to Fermilab, he tried to salvage the situation, demanding that his American colleagues publish their neutral-currents results immediately. Cline and Mann were so peeved by Rubbia's attitude—they objected to being ordered around as though they were graduate students—that, instead, they got nervous and decided to redo their experiment. With more data, and under brutal pressure from the Fermilab management and the physics community to come up with a definitive answer, they concluded prematurely that they did not have evidence for neutral currents. When they told Rubbia, he believed them. He set about spreading the word at CERN that neutral currents were nonexistent and gleefully (as Sulak put it) tried to convince the management to start an investigation into how the Gargamelle people could have propagated such a grievous error. ("If you can't be a giant," Martin Perl, a California physicist once told me, "you want to be the giant killer. That's the name of the game.") With Rubbia's machinations rolling merrily along, the HPW physicists continued refining the second analysis, and realized that they did, indeed, have neutral currents.

By the time it was over, Rubbia and his colleagues were the butt of one of the more biting jokes in the physics community: They were said to have discovered alternating neutral currents. (Dave Cline, whose initials are D.C., picked up the nickname "A.C.D.C.") If Rubbia's backroom politicking accomplished anything, it was to help muddle the situation so thoroughly that none of the Gargamelle people ever got the Nobel Prize their work seemed to deserve.

Within a year, the neutral-currents episode had faded in everyone's memory, giving way to excitement over the discovery of a particle that would become known as the J/psi. And Rubbia, in response, was on his way to his third and last critical mistake of the decade.

On November 11, 1974, Sam Ting at Brookhaven, a laboratory on Long Island, announced that he had discovered the J particle—J being the Chinese character for Ting, and Ting being a man who is not known for his modesty. That same day, Burt Richter at the Stanford Linear Accelerator Center, known as SLAC, announced that he had discovered the psi particle. The two particles were at the same energy, 3.1 GeV, in two different types of machines. Ting's experiment was on the butt end of a beam of protons. Richter's was on a machine colliding electrons and positrons, their antimatter* counterparts. Ting in fact had had the effect for over a month, but had not trusted it. Richter had had the effect for one day. Ting happened to be visiting Stanford, and heard Richter's news and told him his. They called it the November Revolution.

Until the discovery of the J/psi, as it came to be called on the east

---

*A particle of antimatter has the same mass as its matter counterpart, but its electric charge and other quantum numbers are opposite.

coast, or the psi/J, as it is known on the west, physicists knew of
only three species of quarks, which they term "flavors." (Flavor is
a characteristic of quarks similar to electric charge for protons and
electrons.) The two lightest were called "up" and "down" and are
the constituents of protons and neutrons. The third were "strange"
quarks, called strange because when the first particles containing
such creatures were discovered in 1947, nobody knew what to make
of them. But now there was a fourth. The J/psi seemed to be a
particle made up of a "charmed" quark (Sheldon Glashow, along
with Luciano Maiani and John Iliopoulos, had postulated charmed
quarks and whimsically named them ten years earlier), and an
"anticharm" quark, although they didn't finally prove this for another couple of years. Glashow and his colleagues had written
charm into existence to close some loopholes in the Weinberg–
Salam model of the electroweak interactions. Its existence fit into
the theory perfectly.

Six months before the J/psi discovery, Rubbia had announced at a
London conference that his HPW experiment had found a couple of
events—referred to as dimuon events because they each contained
two muons—which, he speculated, could be attributed to a new
hadronic quantum number. In other words, a new quark. In reality
they were evidence for charmed quarks, but he did not speculate
loudly enough. When Ting and Richter broke their news, Rubbia
knew what he had missed. He immediately blamed his Wisconsin
colleagues for hiding other dimuon events from him, and hence
preventing him from claiming the discovery, which was rightfully
his. In fact, his colleagues had hidden several events that they
believed to be merely background events,* but which they feared
Rubbia might misconstrue and consider sufficient evidence for the
discovery.

Ting and Richter won the 1976 Nobel Prize.

Having missed one Nobel-caliber discovery, Rubbia, Cline, and
Mann took to pushing one of their own candidates with increased
vigor. It had begun while the neutral-current affair was in full bloom.
In the autumn of 1973, the HPW physicists found that their neutrino
data did not fit the theoretical speculations. The going theory of
quarks and the strong force made fairly precise predictions about
how neutrinos and antineutrinos would interact with the quarks in
protons. The predictions came in the guise of spectra known as

---

*"Background" events are those that are caused by mundane, unremarkable physics.

y-distributions, and the HPW data seemed to contradict the theory in the region of "high-y."

If this anomaly were real, it required what one journal called "major upheavals in the underlying theory," and even before Rubbia and his colleagues officially claimed that the anomaly existed, the theorists jumped on it. Rubbia brought the preliminary news to Harvard, and Glashow cooked up a theory, with a young theorist named Alvaro de Rújula, in which they introduced new quarks, this time called fancy quarks. They then propagated this theory as though they had no idea that experimental evidence had been found to support it, in a somewhat halfhearted attempt to appear more clairvoyant than was the case. If Glashow thought a particular piece of physics was worth paying attention to, his theorist colleagues assumed he knew what he was doing. They started referring to the excess events as the high-y anomaly and churning out interpretive papers with a fury. The fever spread from Fermilab, Harvard, and Wisconsin in concentric waves.

With the discovery of the J/psi, Rubbia's two colleagues, Cline and Mann of Penn, became increasingly excited about the anomaly. If Richter and Ting could find the J/psi when they weren't even looking, maybe there were other gold mines hanging about. Cline was a radical out of Berkeley with shoulder-length hair who was considered dangerously imaginative ("radical, crazy people," is the way Cline described himself and Rubbia). Mann, however, had a reputation as a down-to-earth scientist who could be counted on to do serious physics. But they were all excited about the high-y, and the theorists backed their optimism all the way.

By this time, Rubbia had almost completely split from the HPW collaboration, angry at his colleagues over both the neutral-current fiasco and the dimuons. He was spending most of his time on a new experiment at Brookhaven, or on his latest ISR experiment, or designing yet another experiment for CERN's next accelerator, which was scheduled to come on in two years. When Rubbia did come to Fermilab—about once a month—Cline and Mann would show him their analysis over dinner or lunch. Rubbia would then take the latest results, with Cline's interpretations, back to Europe. The postdocs and grad students on the experiment, who were doing all the work, and who had grave suspicions about the validity of the high-y anomaly, would never even know he had been there. The three group leaders would then lecture around the world, propounding the extraordinary possibilities of the high-y.

When the collaboration wrote up journal papers on their results, they played down the wild interpretations much more than they did in talks. The papers were tame by comparison. Before a paper could be written, the group had to meet en masse to discuss it. The young post-docs, who worked on the machine and the analysis day and night, would argue that their data had serious problems, that the data weren't nearly as clearcut as Rubbia, Cline, or Mann seemed to think. (The fact that the papers had to be talked over in the group meetings actually helped to salvage Rubbia's reputation. Years afterwards, physicists would say, as Ñick Samios did, "Carlo is imaginative. He likes to think of new things. Watch what he publishes, not what he talks about.")

In the fall of 1976, the HPW collaboration—with Rubbia only an occasional presence—cranked up experiment e310, a rebuilt version of the first experiment, and reconfirmed the existence of the high-y anomaly. The collaboration also came up with a new anomaly, called the trimuons, which Cline postulated could be indicative of a brand-new heavy lepton (which would be like an electron, only more massive) that he called the L particle. The high-y itself had now been going around the physics community for three years, and theorists were still writing papers; meanwhile grad students were doing problem sets on it and writing Ph.D. theses to try to explain it.

A year later, Jack Steinberger, Rubbia's old adviser and colleague, successfully killed both anomalies. Steinberger had begun an experiment on the brand-new 400 GeV Proton Synchrotron at CERN—a souped-up version of the Fermilab machine. Whereas Rubbia, Cline, and Mann had been basing their interpretations on data from a handful of bizarre events that really could mean almost anything—rather like claiming a die is loaded after rolling it only three times—Steinberger had a gleaming new detector that catalogued thousands of neutrino interactions. The HPW data had been almost right, but the interpretations turned out to be much less so. What HPW had claimed were anomalies unexplained by any theory were in fact easily explicable by the standard model.

Steinberger's paper, published in July 1977, was entitled "Is There a High-y Anomaly?" The title alone was considered racy and aggressive for the staid journals of physics. Instead of publishing the paper in the European *Physics Letters,* as all CERN physicists did, Steinberger sent it to *Physics Review Letters,* the American journal. He explained that since that was where the high-y papers were published, that was where the correction should be. Physicists who knew both Rubbia and Steinberger, and knew that they had worked together in the sixties and split not the best of friends, suggested that Steinberger's motive was revenge.

As for the reputations of the individuals involved, the physics community had no great delusions about what had happened. Nick Samios wrapped it up this way: "The problem was that Carlo is very imaginative. This was compounded by Mr. Cline, who's off the wall. The way I would put it, the third man of that contingent, Mr. Mann, should have been the sobering influence. He turned out not to be. So, as a result, there was no one who tried to take the conservative point of view. They all went out for glory." And Robert Wilson said, "They were trying to play all the games, and playing them to the hilt. They just didn't do it well."

Years later, Rubbia himself summed it up in his own inimitable style. He told one of his close colleagues: "Jack Steinberger took all the fun out of neutrino physics."

2.

FOOTNOTES TO HISTORY

"This generation of high-energy physicists could also have done very well in
the retail garment trade."   —Marty Perl, discoverer of the tau particle

For the eight or so years before 1973, nothing much had
happened in weak interactions other than a lot of subtle measuring
of various arcane subatomic quantities. Work that physicists refer
to, some with slight disdain, as bread-and-butter physics. Then suddenly the neutral currents came along in 1973 and the J/psi a year
later, which firmly established the concept of quarks, perhaps once
and for all. It was maybe the most exciting two years of physics
ever.

All physicists knew what had to be done next. The neutral-
currents discovery meant that the Z existed and most likely the W's
as well. Nobody doubted it. It also meant that the Weinberg–Salam
model for wedding the electromagnetic and weak forces was almost
assuredly correct. But physicists were still skeptical. It could be that
almost any variation on the theory was the correct one, and the only
way to find out was to make the W's and Z's and see what they
looked like. The Weinberg–Salam model even predicted the mass of
the particles, and the measurements from the neutral-current experiments provided the numbers—which at the time pointed to
masses of between 50 and 100 GeV.

It seemed that for the first time physicists knew exactly where
to look for the W's and Z's, and what to look for. But they needed
an accelerator powerful enough to do it—a machine that could concentrate enough energy into one infinitesimal point in the universe
so that if the intermediate vector bosons existed, they would be
created there. And although everybody seemed to be building accelerators, none of them would be able to do the job.

At CERN, the Europeans were building the Super Proton Synchrotron, a beautiful 400 GeV machine, but because it was a fixed-

target machine and not a collider, only about 30 GeV of that energy would go to creating new particles.* It wouldn't even be in the ballpark for W and Z energies.

At Fermilab, Robert Wilson had finished his main ring with $25 million to spare. So as not to lose his accelerator physicists, he immediately put them to work on a new machine that could eventually be hung above the first one and, with superconducting technology, get protons spinning at 1 TeV. Wilson called it the Energy Doubler. But it was also a fixed-target machine; intriguing as it would be with superconducting technology, it, too, would be in no danger of finding the W or Z.

At Brookhaven, they had plans for a superconducting machine known as Isabelle that had been conceived back in 1970. It would collide protons and protons at 200 GeV in each beam.† But although that should be enough for making W's and Z's, the Brookhaven physicists had yet to nail down either the superconducting technology or the money from the government needed to build it.

Since the Vietnam War had ended, the federal government had, at least on paper, spent no construction money for new physics facilities. The only new machine that had been built—an electron-positron collider called SPEAR at the Stanford Linear Accelerator Center (SLAC)—had been pulled off solely through the graces of a comptroller of the Atomic Energy Commission who had been able to get the machine classified as a piece of equipment rather than a construction project.

In the summer of 1975, the HPW collaboration at Fermilab had fallen apart. The grad students and post docs were still doing physics, but the upper echelon—Cline, Rubbia, and Mann—had fractured. "I was hardly speaking to Carlo," Dave Cline recalled, "and I was thinking, I don't want to work with this crazy guy no more." Cline was at SLAC, trying to forget Fermilab.

*Conventional accelerators are fixed-target machines, in which particles are accelerated and then smashed against a target that is stationary. Only a small fraction of the energy in the incoming particle, what is called the center of mass energy, actually goes into creating new particles—in the same way that when a moving billiard ball strikes a stationary one, most of the energy goes into the subsequent forward motion of the two balls, and not into destroying either one.

†In colliding-beam machines, which represented an economical advance in atom smasher construction, particles are accelerated in opposite directions and then smashed head on. The two incoming particles may totally annihilate each other in the collision, in which case all the energy of the beams goes into creating new particles. The center of mass energy is then equal to the sum of the energy in the two beams.

Cline is from Kansas. He doesn't say much about his background other than that his parents were middle-class and that he spent a lot of time on the streets. And that he had always been intrigued by philosophy. Given the slightest opportunity, he will talk with wonder about the nature of discovery or the difference between religion and science. He moved into physics because he found elementary particles the closest thing to philosophy on which anyone could make any real progress. He did his Ph.D. at Berkeley in the sixties and began to cultivate a reputation as the tassle on the lunatic fringe of physics. (He once visited Rubbia in Geneva in 1972 dressed like Buffalo Bill, with a white suit and a white cowboy hat, and his shoulder-length hair.) Cline traveled so much in the course of his physics work that, as rumor had it, his graduate students once calculated he had an average speed of forty miles an hour. (Rubbia's flying, on the other hand, had already earned him the sobriquet of the "Al Italia professor." As a physicist, Cline had an almost clairvoyant ability to pick up the signals of worthy new ideas.

While they were at SLAC, Cline and Burt Richter wrote a letter to Wilson at Fermilab suggesting that with a little ingenuity they could collide protons from Fermilab's main ring with protons from the Energy Doubler and end up with the kind of energy needed for creating the vector bosons. A proposal had already been circulating around Fermilab to do the same thing, but Wilson had ignored it. He had doubted that the main ring was built well enough for colliding beams, since it would have to keep two beams circulating for hours, or even days, to be efficient. Cline and Richter simply suggested that the scheme would work provided they used the main ring at about a tenth of its potential power. Nobody else had thought of this. Although Richter was more interested in electron-positron colliders than proton-proton, his name gave the proposal more credibility than Cline's alone could provide.

In July, 1975, Rubbia also wrote to Wilson suggesting proton-proton collisions at Fermilab. Wilson immediately saw what the three physicists had in mind: a quick and dirty and very, very powerful machine that could beat Brookhaven's Isabelle by several years to the choicest piece of physics of the decade. Wilson suggested they hold a workshop the following January, and Cline, who helped organize it, made sure that Carlo Rubbia would be there.

Nobody invited Peter McIntyre to the January workshop, but he heard about it and went anyway. McIntyre was one of Rubbia's disciples: a very bright, very young post-doc at Harvard who had

been working at one of Rubbia's ISR experiments. McIntyre was
fifteen when he left high school in rural Florida and went off to the
University of Chicago. He finished his undergraduate degree at age
nineteen, and received his Ph.D at age twenty-four, in 1972. He is
tall, lanky, and aggressive, his Florida accent not much diminished
after nine years up north, his ambition not diminished at all. He
went to work for Rubbia because Rubbia, he decided, was probably
the brightest man at CERN, and, in his words, a swashbuckler.
While he was there, McIntyre played with the idea of making a
proton-antiproton collider.

The workshop was held in the auditorium at Fermilab. Cline began
with his calculations detailing what exactly they needed in an accelerator in order to find the intermediate vector bosons based on
the Weinberg–Salam model. Then Rubbia explained how proton-
proton collisions between the main ring and the Energy Doubler
would fit the bill.

Then McIntyre talked. To an audience of maybe thirty of the
best accelerator physicists in America, he stated how they might
collect antiprotons and put them in the same ring as the protons—
because the same magnets that push protons in one direction will
push antiprotons in the other—then spin them both up to speed
and collide them. And all in the same beam tube, so that they
didn't need to worry about the problems of the two rings. He explained that there was work being done in the Soviet Union which
would allow them to collect enough antiprotons, and he even
worked out the numbers on the board and showed them what they
might expect. And he showed them that they ought to be able to
nail the intermediate vector bosons that way—if they wanted to
take the chance.

But McIntyre was not an accelerator physicist. He had never
even read a book on the subject, let alone built an accelerator with
his bare hands like half the physicists sitting in the audience. And
what's more, he was a kid. They laughed at him. "Where are you
going to get these antiprotons?" they said. "From another galaxy?"

Still, his audience wasn't totally unreceptive. When they talked
about it later, they saw that within McIntyre's scheme lay a shortcut
that Fermilab might take to beat out Isabelle. Many of them would
readily acknowledge that Isabelle represented the avant-garde of
American physics, the machine of the future. But that simply didn't
matter if they could get to the physics first. "We would be very much

in the running," Wilson said, looking back. "That's what we were talking about."

The night after the workshop, Rubbia flew back to Boston with McIntyre. The antiproton idea had him fired up. "We had this euphoric discussion all the way back to Harvard," McIntyre remembered. "You know how Carlo works. Once you get going on a productive piece of physics, it's sort of creativity at a kilohertz pace." Nobody learned faster than Rubbia. He could get from a point of total ignorance to near expertise almost instantaneously; nothing had to be repeated. When the plane landed in Boston, Rubbia already had McIntyre outdistanced on the mechanics of proton-antiproton collisions and he was working out improvements on the scheme that McIntyre had never even considered. Rubbia decided to work with him, in much the same way that a cat decides to work with a canary.

A few days later, Rubbia, who had originally proposed a proton-proton experiment, called Cline to say, "Look, in this proton-antiproton thing we've hit on a great new idea. We should do it because it's unique. Let other people do the proton-proton."

The innovation in the antiproton scheme was not the making of antiprotons, but amassing enough of them once they'd been made. Making them was easy. You just take an accelerator and bombard protons onto a target. Thousands of secondary particles will be created in the collisions, including a small percentage of antiprotons. You then use magnetic fields to skim away the antiprotons from the other debris before they have a chance to collide with protons and annihilate each other. Unfortunately, your antiprotons will have a wide range of random motion and no more than a few thousand at a time will fit into a beam tube to be accelerated. The resulting problem is something like trying to funnel an angry swarm of bees into the barrel of a rifle. Proton-antiproton collisions had been raised as a possibility for Isabelle in the early seventies, but the technology needed to gather sufficient antiprotons to make it practicable was not ready. What McIntyre brought to the Fermilab meeting was the news that the technology—known as cooling—had now arrived, or at least was on the way. "The word was dropped at this meeting," Wilson said, "that you could cool beams."

Beam cooling is a method of reducing the random motion of antiprotons (called cooling because the smaller the random motion, the lower, in effect, the temperature of the particles), so that trillions of them can be rammed into beam tubes, accelerated, and collided. "We had to go from a hundred antiprotons in old experiments," said Rubbia later, "to a hundred billion antiprotons in nice beams every morning with breakfast."

Two potential techniques existed. Gersh Budker, a brilliant, funny, iconoclastic Russian Jew, had invented one, called electron cooling, and had tried it in his laboratory in Siberia with little luck. But by 1975, McIntyre had heard rumors that the technique was looking promising. The other technique was known as stochastic cooling, courtesy of Simon van der Meer, an equally brilliant CERN engineer. He had concocted it in 1968 as a possible way of increasing the density of the beams in the ISR. Later, he let it drop, saying that it was "too far-fetched." Since then the CERN accelerator physicists had regenerated it, tested it on the ISR, then shelved it again, pronouncing it interesting but not particularly useful.

Rubbia returned to CERN and spent the rest of the winter of 1976 in the library and at the computer terminals absorbing everything he could on both cooling techniques. Rubbia's history on proton-antiproton collisions—what some physicists refer to as pbarp (pronounced pea bar pea) and others as p-pbar—went back a long way. In the mid-sixties he had worked on the idea of building such a collider in Italy for a machine with 80 GeV in each beam. A few years later he had proposed pbarp for the CERN Super Proton Synchrotron, which at the time was still four years away from even being a hole in the ground. Rubbia had known the great Austrian scientist Bruno Touschek, who had fathered electron-positron machines in Europe. He had met Budker over dinner at Gerard O'Neill's house in Princeton. O'Neill was a Princeton physicist often credited with coming up with the original idea of colliding beams back in the fifties. And Budker had written about building a pbarp machine at his laboratory in Siberia back in 1965, and had been plugging at it off and on ever since. "He was the man who really had all these ideas before anybody," said Rubbia, who still keeps a photo of Budker on the wall of his office.

By March, Rubbia knew as much about cooling as any accelerator physicist. Even physicists who had considered him nothing more than a brighter-than-average flake were impressed. "Carlo had read the whole literature in the accelerator field and knew it," said one. "Most of the people in the field hadn't done that."

With Cline and McIntyre, Rubbia mailed a paper to *Physical*

Review Letters, in which they outlined a scheme to search for the W particle by adding an antiproton source to either the big accelerator at Fermilab or the similar one at CERN. The goal, they wrote, was "to transform a conventional 400 GeV accelerator into a p-pbar colliding beam facility with 800 GeV in the center of mass. . . ." The journal rejected the paper. Brutally. One referee said it was "entirely inappropriate" to publish. Cline felt that the journal—which is located at Brookhaven—rejected it because it represented a head-on confrontation with Isabelle.

The trio next submitted two proposals to the Fermilab management: one for a cooling apparatus to produce antiprotons, and one for the detector to do the physics should the scheme go through. Rubbia had written the bulk of the proposal in Geneva and had mailed it piecemeal to Cline in Wisconsin. Cline managed to get it to Fermilab and to hand it to Wilson only thirty minutes before they were to report on the scheme to a meeting of the lab's board of trustees. Wilson was a bit put off. "Now, you don't do that to a director," McIntyre noted. This sparked what McIntyre calls his political education. Cline calls it the Wednsday night massacre. Rubbia prefers not to talk about it.

Robert Wilson was a cowboy from Wyoming. He had learned to build accelerators by hand at Berkeley under E. O. Lawrence, who had more or less created the accelerator business in the United States. Wilson had worked on the Manhattan Project. He had built Fermilab from the ground up into a monument of science and, indeed, of art. The design of the buildings and equipment are both architecturally breathtaking. Wilson was bold, outspoken, piratical, and sometimes cruel. He was known for his temper, and he had gotten jobs done even if it meant insulting his peers. He neither respected nor liked Rubbia. The neutral-currents affair hadn't endeared him to Rubbia, and, at the time, the lab was in the midst of the strife over Rubbia's high-y anomaly.

Wilson had been defending Cline and Rubbia over the high-y anomaly and the dimuons, and had gotten them more running time to improve their statistics and try to come up with something firm. But he was getting tired of it. Around the lab, he would hear doubts expressed about the high-y. From outside physicists, he would hear the wild interpretations that Cline and Rubbia had pushed in seminars. And then he would hear that instead of working on the experiment, Cline and Rubbia had been flying around talking about proton-antiproton colliders.

"This experiment," Wilson recalled, getting angry even a decade later, "was our main shot, and they were the main people, and here they were flying about and talking about this other scheme. Plus I was giving them a run which was not approved, and this was in some measure a violation of what had been recommended. And, believe me, my neck was way out, and here those guys were all talking about something brand new. And, oh boy, I was pretty angry."

At the trustees' meeting, Wilson was civil. Later, after considerable drinking over dinner, he got the two physicists alone in his office. (McIntyre waited outside, finishing his dessert.) Wilson used, as he put it, "language that you might have thought I brought with me from Wyoming." Among other things, it was then that he called Rubbia and Cline "jet-flying clowns." He told Rubbia he would be no more than "a footnote to history." And he suggested, strongly, that they return to their neutrino experiment and forget about accelerators.

The next morning, Rubbia and Cline met for breakfast at six. Neither had slept. Rubbia said he was finished with Fermilab. He would quit. He was returning to Geneva and Cline should come too. They then drove to Fermilab, and walked into Wilson's office. "We were going to go tell the guy to stick it up his ass," said Cline. But they never got the chance. Overnight, Wilson had had time actually to read the proposal. He liked it, even admired it. He gave the two physicists permission to present the proposal to the Program Advisory Committee, which would meet a month later. The members of this committee were physicists themselves, and they were responsible for suggesting what the director should do with his lab. Rubbia decided not to quit. Not yet.

When the committee met in June, they had three proposals to decide upon. One was the proton-antiproton, and the other two were different schemes to collide protons with protons. Of those two, one was supported by Leon Lederman and backed by a strong contingent of respected physicists. The other was being proposed by a large coalition of Fermilab in-house physicists. By contrast, the antiproton scheme was supported by a group of mavericks. For diplomatic reasons, at least, Wilson and his committee adopted an attitude of benign neglect. They would work on all three schemes —a little.

Rubbia was angered by this arrangement, but he wasn't finished. He tried to convince Harvard to throw in money so that he could work on the project at Fermilab with a strong Harvard presence that would help him to assert his strength in the project. The Harvard physicists supported Rubbia's idea, but Rubbia was already using 40 percent of the total high-energy physics budget, and the department had no money for new schemes. When Harvard appealed to the Department of Energy for more funds to back Rubbia, they were refused. Rubbia then tried to compromise by getting a written agreement from Wilson that should an antiproton source be built at Fermilab, Rubbia, Cline, and McIntyre would be put in charge of it. But Wilson refused. Rubbia summed up the situation candidly: "Wilson didn't like me, but he liked the idea. In fact, he told me very clearly that he would throw me out and do it anyway. It was not gentlemanly behavior." And he added, "If the antipathy was for the neutral currents, it should not have been for me, it should have been for the other two characters."

Rubbia had already begun pushing the pbarp idea at CERN.* (Cline and McIntyre had a friend at CERN who regularly Telexed them the latest developments. In early June he wrote, "An informed source has told me that Rubbia has dropped out of your Fermilab proposal. He is pushing the proposal with great vigor here." This agent-in-place also served another purpose: "I am assembling a dossier of 'classified' ISR documents on beam cooling, which seems to be the crux of the whole business," he wrote.) By now, Rubbia was concentrating almost exclusively on CERN. He was motivated by revenge as much as anything: the desire to prove to the American physics community, specifically to Wilson, that they had missed the chance of a lifetime.

"Of course," Rubbia said, when I asked him about this nine years later, "that's a fair response. There's nothing bad about it. It was according to the Marquess of Queensberry's rules. [Wilson] ended up k.o.'ed."

---

*John LoSecco, a physicist now at Notre Dame, told me that he saw Rubbia after the first rejection at Fermilab, and suggested that he try pbarp at CERN. Rubbia supposedly replied that, for lots of reasons, it would never work at CERN. "I said he could get CERN moving fast on the project," LoSecco told me. "He threw me out of his office. He said, 'Boy, that's a stupid idea.' Two weeks later, CERN was moving on pbarp."

3.

THE SELLING OF
THE COLLIDER

"From [Ernest] Lawrence onward, if you really wanted to become famous in our field, you had to get yourself the biggest and most powerful accelerator."
—Carlo Rubbia

Rubbia heard about the discovery of the J/psi while he was at Harvard. He glanced at a copy of the preprint, turned to one of his graduate students, and said, "Take a look at this bombshell."

The news hit CERN a little harder. After the story broke, the CERN physicists called an emergency meeting in the theory division conference room. They all came. Even the management came. One of Richter's team from SLAC flew into town and described what it was they had seen. Then, as Peter McIntyre remembered it, "There was a sort of collective wringing of hands that went on for probably an hour. Some of them were crying, literally crying." It was the first meeting of what Leon Lederman would later call the "I Missed the J/psi Club."

The physicists of CERN sat in the lecture hall and told each other how they had lost the discovery of the decade. The Intersecting Storage Ring at CERN had been making five times as many J/psi's a day as the machine at SLAC, and yet the CERN physicists had managed to miss them all. Lederman and Luigi DiLella, an Italian physicist, had had an experiment on the ISR in which they had caught the beginning of the peak of the J/psi, but they had so much background they decided they would just ignore everything with an energy of less than 3.2 GeV. The J/psi was at 3.1 GeV. DiLella then put together a second experiment that at the time of the announcement actually had 11 J/psi's in the data, but no one had realized it. A British group had set up when the ISR came on, stayed for a year and a half while the machine was still getting debugged, then packed up their experiment and gone home. If they had stayed around . . .

There was even a Belgian physicist who at the meeting read a two-year-old memo written when the experiment they had proposed had been rejected. The local theorists had said then that they didn't expect any new particles in that energy range, and so there was no reason to look. One of the things the memo said was "We think that it would be infinitely regrettable, not only for us, but also for CERN, and for the European scientific community, that an experiment which opens a new field of research would be definitely forbidden when a similar experiment, although less rich in its possibilities, is accepted and encouraged at BNL [Brookhaven National Laboratory]." That "similar experiment" was being run by Sam Ting. In fact, Ting had first suggested doing the experiment at CERN, but the management had told him not to even bother writing a proposal because there was already a European proposition for the same experiment.

Yes, it was all "infinitely regrettable" to the CERN physicists that day. The lab had been created in the early fifties to restore European excellence in a field that had been unquestionably theirs until World War II, and to compete with the huge American labs that were then springing up. CERN was a beautiful idea, intended to help pull the continent back together as well as to restore pride in European science. But in twenty years they had done nothing but mediocre physics. From the first neutrino beam that was short of neutrinos (the muon neutrino was eventually discovered at Brookhaven, even though the CERN machine was on line a year earlier), through the sixties when every major discovery was made in the States, the only things CERN discovered were a W particle that was not a W, and a particle called the A-2 that they believed was actually two particles when it wasn't. In the seventies they had discovered neutral currents, but between Rubbia's political infighting and the uncertainty at Gargamelle, that affair had been so botched that it somehow seemed more of a failure than anything else. ("The Europeans have a particular habit of always stressing the back side of things," said Leon Van Hove, one of the director generals of CERN during that time.)

When Rubbia started pitching the proton-antiproton collider idea at CERN in early 1976, he played on that chord like he was Horowitz. The following fall, when Richter and Ting won the Nobel Prize (in twenty-five years CERN had yet to win a Nobel Prize), the European physics community was on its way to developing a massive inferiority complex. Then in the summer of 1977 Lederman announced at

Budapest that he had discovered the fifth quark—in the guise of a particle known as the upsilon—at Fermilab. All CERN had at that time was Jack Steinberger saying that there was no high-y anomaly. And you don't win Nobel prizes for saying that something doesn't exist. "I met Jack at Budapest," Lederman recalled, "and he was really pissed off because he had the goods on them—no high-y anomaly—but I had the upsilon."

Again the CERN physicists looked into their past and realized that they had rejected as being overly grandiose two experiments that might have found the upsilon. They even had one experiment that seemed to have found the upsilon during a preliminary run. But the physicist who reported it was criticized so vituperatively by his colleagues that he subsequently reported that there was no signal.

The Europeans couldn't quite understand it. They had the ISR, the most powerful machine in the world by a factor of thirty, a beautiful machine, the only proton-proton collider ever built, and it worked like a dream, better than they had ever believed possible. Yet the physics that came out of it had been a public relations disaster. They had done some beautiful bread-and-butter physics. But that was it. "Nobody will ever go to Stockholm for any of the experiments done at the ISR," Rubbia had said.

Even worse, the one really notable discovery at CERN in those years, the neutral currents, was made at the twelve-year-old Proton Synchrotron with an outmoded bubble chamber, Gargamelle. Yes, CERN had an inferiority complex, and Rubbia played on it. He was convinced he had the cure: the proton-antiproton collider.

All Rubbia had to do was to convince the entire European physics community and, more importantly, the CERN management, of the practicability of his idea. Both of these parties had a reputation for being conservative, however. And many felt that this conservatism was the reason that CERN's physics program had been so overwhelmingly mediocre. Whenever the management was presented with a plan for an ambitious machine or detector, they would talk it over for a couple of years and then approve it in a diluted version, one step at a time, until finally somebody else beat them to whatever the original goal had been.

But now they were all too aware that they had a problem and that the situation wasn't going to improve in the near future. In the fall of 1976, the new machine, the Super Proton Synchrotron (SPS), was due to come on line; but it was, as everyone knew, just a classy version of the Fermilab main ring, and it was coming on five years later than the American machine. The two machines had been conceived roughly at the same time, but the Europeans had opted to first build the ISR and had then spent three or four years haggling over which of the member nation of CERN would get the SPS. ("I myself went to several of the sites," said Don Perkins. "I remember going to a site near Madrid. It was a plain in the mountains, and on the mountain there was this great cross erected by the Fascists. It was to commemorate the ones who died during the Civil War, the ones on Franco's side. And I said, 'I'll tell you one thing. If I know anything about CERN, if you want the machine to come here, the first thing you have to do is take down that bloody cross, because nobody is going to have an international laboratory with that political symbol on the mountain. No way.' ") It wasn't until they realized that they could construct the thing sixty meters underground, and therefore put it in suburban Geneva—and save $100 million in the process—that they got around to building it. By that time, they had a machine that would do only competent but relatively boring bread-and-butter physics.

By 1976, however, the CERN management was surprisingly amenable to any idea that might break this cycle of failure. They had two healthy inducements. First, like their American counterparts, they knew that the neutral-currents work had pinned down the masses of the W and Z, and that anybody who had the right accelerator could go out and claim the discovery and the Nobel Prize that would go with it. Second, they had a problem with money. Unlike the Americans, they had too much of it. The lab had plans in the works for its eventual saviour, an enormous half-billion-dollar machine called the Large Electron Positron collider (LEP), which would be sixteen miles in circumference and would collide electrons and positrons at an energy high enough to create the W and Z. But they wouldn't have the technology for this ready until 1981 or 1982, which in turn meant that the machine wouldn't be doing physics for at least another five years after that.

LEP in some sense was also a child of the J/psi discovery. When Richter's team found not only the psi but a cascade of different states of the psi, physicists realized that these electron-positron collisions had a lot to be said for them, and started clamoring for their own machines. The Europeans had had a couple of e+e− machines, as they were called, in Frascati, Italy, and they had done good physics. But when those amazing results came out of SLAC, everyone wanted to build bigger ones. The Germans started planning an electron-positron machine at a laboratory in Hamburg, the British started planning one for the Rutherford Laboratory near Oxford, and then canceled it when the German machine got off first. And CERN decided to take on LEP.*

Since the lab worked on a five-year budget plan, there would be a five-year gap between the time the SPS was finished and LEP was started. If they had no intervening project, the European governments that provided funds for CERN might cut their budget, in which case they would have trouble raising it again for LEP. "Everybody felt there was a hell of a lot of money there," Rubbia recalled, "because CERN had finished building the SPS, and budgets had not come down. Of the construction power that you had to build the SPS, half was now running it, the other half was available for other things. That had to be kept in good standing for the day LEP would start. A project was needed."

A handful of projects were floating around to take up the slack in the budget. Many of the ISR veterans were pushing to convert their machine into a superconducting one, which, although it would be smaller than Isabelle, could be built more quickly. Another contingent wanted to build an electron ring, and collide the electrons with the protons in the SPS. Both machines would have the power to make the intermediate vector bosons, the W and the Z, but the upgraded ISR seemed to be the number one contender.

Then Rubbia came along pitching the antiproton idea in early 1976. At the time, he may have been the only person at CERN who believed it could work. The other physicists didn't understand cooling and they didn't trust it. The idea looked too risky. The machine people said so. And they had the other ideas, which looked so much more certain, and had much more trustworthy physicists behind them.

Rubbia's reputation was the immediate problem. Virtually every major physicist at CERN had worked with him at some time or another and had fallen out with him and moved on. He was not the type of man you wanted to hand over the future of your laboratory to. There was a joke that went around CERN for years about the kid

*The original suggestion again came separately from Richter and Rubbia in 1975. As late as the winter of 1976, Rubbia was still publishing papers saying that the only way to test the electroweak model would be with electrons and positrons.

who asked for a cowboy outfit for Christmas and who got Rubbia's group. Then, as one physicist put it, he was coming off "the magnificent blunder of the alternating neutral currents." (Another young physicist told me that "Rubbia was just the sort of a name that you laughed at because he didn't know what he was doing.")

They also knew that Rubbia was full of ideas, crazy, Utopian ideas, so many that he usually forgot them himself. This looked like another one. Then, if they stopped to consider it, they realized that turning the SPS into a collider would mean shutting down the machine for a year while they refit it to run as a storage ring. The SPS had just been completed, and a lot of physicists wanted to do good fixed-target physics on it. They weren't overjoyed by the thought of their experiments being shut down for a year just to make Carlo Rubbia happy.

The idea seemed just too radical for the conservative CERN management, no matter how much they might want to change. Nevertheless, Rubbia went to work selling the collider with his inimitable frenzy. He kicked down doors, he forced people to listen to him, he argued and harangued and charmed—"He really lathers you up," said one American—and seduced, and bullied. He held workshops, wrote long lists of the people who might be interested, then badgered them to come. When he got them there, he put on his show and sold them pbarp like it was Dr. Rubbia's Magical Miracle Cure. He said later that maybe the reason his show had bombed in the States was simply his "failure to sell my dream."

One of Rubbia's most effective sales techniques was to let CERN know that he had another bidder. Through McIntyre and Cline, he had continued pushing at Fermilab, and he kept it no secret. "Carlo had a hand in both labs," Wilson said later. "And he was using one to amplify the other. I say that is fair game. He did it masterfully, whatever the hell he was doing."

At Fermilab, Wilson's decision to adopt an official attitude of benign neglect on the pbarp project meant that the laboratory's first priority would be to go on building the Energy Doubler—the 1 TeV superconducting fixed-target machine—as a research and development project. A distant second priority would be to build a colliding-beam facility once the doubler was done—maybe proton-antiproton, maybe proton-proton. Nobody knew.

Either way, Wilson had a political problem to attend to. In 1974, a panel of physicists advising the Atomic Energy Commission (the predecessor to the DOE) said that what the United States physics community needed was a tripartite system to, as they put it, "push the energy frontier forward" on three fronts. In the process it would thwart the competition and keep the politicians happy. They suggested that on the east coast, Brookhaven, on Long Island, would do colliding-beam physics, and this meant building Isabelle. The Midwest would have Fermilab's fixed-target physics, and here they backed Wilson's Energy Doubler. The West would have SLAC, which would do electron-positron physics. Any quick and dirty colliding-beam experiment at Fermilab would not only upset this careful balance among the labs but would make Isabelle, the machine-to-be of the decade, an also-ran. The Washington bureaucrats, as well as the very powerful New York political lobby, had no desire to see that happen.

Isabelle had been officially proposed in 1972 as a 200 GeV on 200 GeV proton-proton collider, with the latest in superconducting technology. By 1976, to the outside world, Isabelle was indeed on its way to becoming the machine of the decade. Within the scientific community, however, rumors were circulating that the Brookhaven people were not having an easy time designing the superconducting magnets they would need to keep their protons orbiting in the ring. But these were only rumors. More troubling was the fact that new measurements on neutral currents pointed toward masses for the intermediate vector bosons that were possibly beyond Isabelle's range. Nevertheless, the Brookhaven physicists were not worried. They could, if they felt that the political currents demanded it, push the machine to 400 GeV on 400 GeV.*

Even with Isabelle as top priority, Wilson would have been in reasonable shape at Fermilab if there had been money to go around. But as it was, high-energy physics in the United States had cash flow problems. In 1975, as one DOE official put it, "the funding for construction at this time was zilch." Wilson had been surviving by insisting that his construction projects were research and development projects and redirecting the necessary money from other funds "at his disposal" from around the laboratory. "What I would do," he said later, "was not pave the roads; not build new buildings; let old buildings fall. Well, if somebody looked into it, I could be in trouble, but you've got to have a little guts."

Even after losing Rubbia to CERN, Wilson still wanted to build

---

*Because protons are made up of three quarks, proton-proton collisions are actually collisions between quarks and quarks. Each quark has, on average, only one third of the energy of a proton.

the pbarp collider. He had seen the CERN SPS, and he had seen the neutrino experiment that Jack Steinberger was building, and he knew that Steinberger was going to "clobber" them. "I had wanted to do pbarp," Wilson recalled. "I was pretty miserable at the time. I felt we could do it and, by God, I was sure we could do it, one way or the other. It was painful."

In Europe, Rubbia made certain that he didn't alienate the CERN management the way he had Wilson. He showed, or developed, political astuteness. Bernard Sadoulet, Rubbia's first recruit on the pbarp project, said, "You could imagine in the beginning that he would step on somebody's feet not knowing what he was doing. Just because of his ambition." But at CERN he never did. Or at least, when he did, he did it knowing full well what he was doing.

At the time, CERN had two director generals. One was Leon Van Hove, a Belgian theorist who had studied under Niels Bohr, had been recruited to the Institute of Advanced Physics at Princeton by J. Robert Oppenheimer, and had run the Max Planck Institute in Munich for three years before coming to CERN in 1974. Van Hove was in charge of research matters. His partner was Sir John Adams, a British technical virtuoso who had helped design CERN's first Proton Synchrotron in the late fifties and had then overseen the construction of the SPS in the seventies. Adams had a flair for administration, and was in charge of all technical matters. "To Van Hove," Sadoulet recalled, "Rubbia said, 'Look, that's beautiful physics. You will be the one responsible for that, remembered for that.' To Adams, who was much more conservative, and very attached to his new machine, he said, 'How will you prevent your budget from being cut if you have no new project before LEP?' "

For Rubbia, Adams was a tougher sell than Van Hove. First of all, Rubbia was suggesting that they deeply modify the SPS, which was Adams's baby, before the machine had even made its first run. Then, Adams was an engineer and Rubbia was a physicist. As Van Hove said later, "Adams had this very strong engineering tradition, that they don't trust these physicists with their fancy ideas." Adams's preference would have been to make an Energy Doubler in the SPS, just like Wilson's; he knew that he could do it better than the Americans could, and he did not care that it would be completed later than theirs. Adams, said Van Hove, "was not a born innovator, but he loved perfection."

In early 1976, Adams and Van Hove decided to go ahead with a crash program to test cooling, called ICE, for Initial Cooling Experiment. It was the perfect compromise for Adams. The lab committed nothing to the future, and they had nothing to lose. If cooling failed, he didn't have to worry about Mr. Rubbia and his fancy ideas. If cooling was a success, then they could talk more, and in the meantime, Adams had an accelerator research project at which to work. "Adams played a fantastic game on the cooling experiment," said Van Hove.

For the first year, nothing visible happened at the lab. Adams freed up a small accelerator, selected a team of engineers, and scavenged magnets from around the lab. In early 1977, the physicists went to work. They hoped to test both stochastic cooling and electron cooling, and maybe use them both. Rubbia was pushing heavily for electron cooling, but it was the Dutchman Simon van der Meer who forced the issue, walking in one morning and, with Dutch equanimity, saying, "I can do everything with stochastic cooling." Rubbia didn't believe him, but van der Meer insisted that he was right. "I'm personally convinced that we might be still checking and tuning the machine if we had gone the route of electron cooling," said Sadoulet eight years later. "Simon had the expertise on his side."

But Sadoulet and the other physicists were still impressed by the fact that Rubbia, an experimental physicist, could speak knowledgeably with accelerator men about accelerator physics. "What we saw," said Sadoulet, "was Carlo not saying silly things, and some of the things he said were even considered very clever by the people in the field, which tells you a little bit about the intellectual abilities of this gentleman."

At that time van der Meer was a fifty-one-year-old engineer, softspoken, modest, the antithesis of Rubbia. But even van der Meer has had a reputation among CERN engineers for tyrannical behavior. He is a workaholic who will sit at a computer console day and night. He is also, as one physicist described him, "a hell of an incredible guy. He is really one of the absolutely great engineering geniuses in the field." Like Rubbia, he was building radio sets when he was fifteen with scavenged parts and his bare hands, and he has never stopped developing. He went on to come up with innovation after innovation in the past twenty-five years of experimental physics. It is easy to find physicists who will say that the only things CERN had that the United States did not were a hell of a lot of money—and Simon van der Meer.

Unlike Rubbia, the CERN management was plugging for stochastic cooling. They knew that if the project worked, they had a Nobel Prize coming, and they liked the idea that it would be led by a CERN physicist relying on a CERN technological innovation. They had no desire to share their first Nobel Prize with Gersh Budker in Siberia.

On October 3, 1977, the United States Congress authorized $10.5 million to begin construction of Isabelle. A few months later, the Office of Management and Budget insisted that Wilson's Energy Doubler must become a construction project or they'd stop it. But there was no construction money. The DOE bureaucrats managed a compromise, but still Wilson was shaken. He felt as though his political support was crumbling beneath him. What made it worse was that by then the physics community was beginning to suspect that Isabelle was in trouble, that their magnets weren't working, and that they were back where they started. They might now have the money to build an accelerator, but they still didn't have the technology.

In late December, CERN began to get results on ICE. Rubbia took to sending Telexes to his Fermilab cohorts telling them of his progress. This was more than information transfer. McIntyre and Cline, along with a small corps of accelerator physicists, were still slugging away at Fermilab, hoping for a breakthrough with the funding. Rubbia was needling them.

One Rubbia letter to McIntyre was dated December 12: "For your enjoyment," Rubbia wrote, "I have joined as well a copy of the CERN *Bulletin* celebrating successful injection of the protons in ICE." A week later, he sent a Telex: "ICE operational again, running nonstop since Friday . . . now proceeding to stochastic cooling studies. . . . So Far So Good." And on December 22: "Last night for the first time ICE has demonstrated stochastic momentum cooling . . . have a very nice holiday."

McIntyre looks back on all this grimly: "He would call me as well. 'We've just done this. Are you listening? We've just gotten this done. . . . Are you listening? Are you listening? How do you feel?'

" 'Like shit, Carlo.' "

Adams had said that if stochastic cooling worked, he would leave the final decision to the physicist, Van Hove. The two together rejected the idea of a superconducting Intersecting Storage Ring for practical reasons: neither Fermilab nor Brookhaven had perfected superconducting magnets yet, and they felt that the time to get involved with it would be when the technology could be simply borrowed from the States. In addition, the ISR consumed a sixth of the CERN budget, and that was money that would be used to build the Large Electron Positron collider. A superconducting ISR might also appear to the European governments as more of a construction project than an upgrade, and thus tempt them to consider it as a replacement for LEP, instead of something with which to idle away the time until LEP. Finally, the superconducting ISR would have simply cost twice as much as Rubbia's pbarp scheme, and would also have taken longer to pull off. The electron-proton collider was already being pursued as an idea for the Deutches Elektronen-Synchrotron Laboratory, DESY, in Hamburg, and would eventually grow into a machine called HERA.

That left only Rubbia's antiproton scheme. Rubbia, consequently, remained realistic about the management's motivation. "I don't think those people took me so seriously," he said later. "I don't think that they thought I was the Salvation Army."

Rubbia had convinced Van Hove easily. Van Hove was a former theorist who saw the possibilities immediately. But he was now left with the task of convincing a dozen European governments that converting the SPS into a proton-antiproton collider was reasonable, even though half the physicists in the world—particularly the American half—did not believe it was. "All they really wanted," Van Hove said, "was not to have to take too much responsibility themselves. There were people who said if it was so good, the Americans would have done it already, and other people turned the same argument around and said since the Americans won't do it, you will not succeed either. But that is normal. We took a risk."

In January, 1978, CERN announced that they would build Rubbia's antiproton-proton collider and along with it the huge electronic detector that would find whatever it was that the colliding-beam machine created. This experiment was to become known as UA1, for Underground Area One.

That same month, President Carter announced that the U.S. budget would include $275 million for the construction of Isabelle. The machine would be upgraded to 400 on 400 GeV. It would be glorious. Someone even sent a senator around to Fermilab to visit Wilson— the cowboy who hated bureaucracy and said that being a member of the establishment didn't correspond to anything he considered respectable—to enlist his support behind the Brookhaven project.

Two weeks later, Wilson mailed his letter of resignation. "The future viability of [Fermilab] is being threatened," he wrote, "because the funding has been below that necessary to operate the existing facilities responsibly. The scheme to increase the proton energy to 1,000 GeV through the application of superconductivity has been confounded by indecisive and subminimal support, as have the modest proposals for intersecting beams. . . ."

Wilson was bluffing. But the bureaucrats called it, and Wilson resigned. "I expected them to come through with additional money," he said. "I wasn't fooling one bit. It wouldn't have taken much to keep that going. But they didn't give me a cent. Not one cent. I didn't want to resign, I must say, but to me there was no alternative." (Wilson received little help from the physicists, many of whom were tired of seeing him channel all the lab's money into the accelerators, short-changing the experiments.)

Leon Lederman replaced Wilson in October, 1978, as director of Fermilab. Lederman is a brilliant experimental physicist with a reputation for having as much desire and imagination as anyone in the field. He had worked on the Brookhaven experiment that had discovered the muon neutrino in 1962, and was famous for claiming that he had missed the J/psi three times—once at Brookhaven, once at the ISR, and once at Fermilab—and had found the upsilon particle twice—the first time mistakenly, leading to the line that Lederman had discovered the "oopsleon." Leon Lederman was charming, witty, clever, and devoted to physics.

When Lederman took the director's chair at Fermilab, he saw that Wilson's benign neglect had deteriorated to what he called the plantation syndrome: "The plantation syndrome is: 'We're free, we're free! The master's gone, we're free! What should we do? Oh my God, you know he's not here? We're free! What shall we do?' " He found a coalition of accelerator physicists at the laboratory who called themselves the underground parameters committee, and who had been meeting for the better part of a year to try to figure out how to save Wilson's Energy Doubler. They had "devastating criticisms," in Lederman's words, of the state of the machine, the lab, and the future.

As one of his first actions as director, Lederman invited the physicists and users of Fermilab to an Armistice Day shootout, on November 11, 1978, in order to establish the future goals of the laboratory. The shootout began at nine in the morning and ended at midnight, during which time they discussed the merits of the various plans that had been struggling along under Wilson's benign neglect. The next morning Lederman had breakfast with his judges. Over bagels and lox they discussed the meeting.

"I came out of that very clear in my mind," Lederman said. "We were going to let CERN win."

The future of the lab would focus on a proton-antiproton collider that would be built in the Energy Doubler once this was finished. The Fermilab collider would be able to hit 2 TeV in the center of mass. But first they would do the doubler right. As Fred Mills, one of the pbarp physicists, put it, "The game was over on Armistice Day, 1978."

A footnote to this story: In 1982, both the government and the physics community agreed that Brookhaven had missed its chance with Isabelle, that the machine would arrive too late. After CERN had shown the first signs that protons on antiprotons could do great physics, Lederman obtained the money he needed to build pbarp at Fermilab. The project was upgraded to $130 million and became a more costly and powerful version of CERN's stochastic-cooling technique. When Isabelle was officially canceled in the fall of 1983, two months after CERN announced the discovery of the Z, the money was freed up for construction to begin at Fermilab. In the fall of 1985, antiproton beams were circulated for the first time in America.

4.

UNDERGROUND AREA ONE

"UA1 was to Carlo what the pyramid was to Rameses IV."
—Don Perkins, professor of physics, Oxford University

On the day that CERN approved the proton-antiproton collider, John Adams, who would spend the last years of his life working to make the project a reality, recited a little poem that now hangs in a frame on the wall of Rubbia's office. "I found that thing particularly insulting," Rubbia recalled years later, "so I decided to put it up there just to remind myself. I was fighting very hard to get his money and his support. When everything was done, and we were all ready to get down to work and start building, we had a party. John came and he stood up and said, 'Well, how about a little poem?', and everybody listened, and everybody laughed, except myself, I guess. It sounds like a *New Yorker* story."

Adams's poem went like this:

> clever simon met a bagman
> leader of a team,
> said clever simon to the bagman,
> i can cool a beam,
>
> bagman said to clever simon,
> i can use this scheme,
> by being first and talking faster,
> bigman i will be.

Rubbia demanded a public apology, and he got it.*

Adams was not alone in his sentiments. The fact that the Europeans would take a risk on Rubbia's project did not mean they were all

---

*In 1984, a week after he received his Nobel Prize, Rubbia gave his Nobel lecture at CERN to an SRO audience of his colleagues. He commenced by pulling his transparencies out of a blue-and-yellow shoulder bag. "This is to keep up my reputation as a bagman," he told them, "although this time the bag has Swedish colors."

that thrilled about ceding to Rubbia the future of the laboratory. When they approved the collider, which would cost in the neighborhood of $60 million,* they also gave the go-ahead on UA1, the largest, most extravagant, most expensive experiment in high-energy physics. It would cost another $20 million by the time it was finished, would be the size of a three-story house and weigh 2,000 tons. Despite the grandeur of the project, however, European physicists were not exactly climbing over each other to be part of the team. For UA1 to succeed, Rubbia's reputation would have to be outweighed by the lure of the physics that might come out of the experiment. So, once again, Rubbia went on the road, this time selling the physics.

His first ally was the thirty-two-year-old Bernard Sadoulet, yet even he says that when he came back to CERN, he had one clear idea: "I did not want to work for Carlo." And when he signed up, he said he'd do it for six months, then take another look.

Sadoulet had grown up in Lyon. His father was an engineer and his mother a philosopher. For a long time, he couldn't decide which to follow. He remembers his father giving him a book when he was twelve or thirteen, entitled, as he remembers it, *Radio Sets Are So Simple,* and that set him on the path toward physics. By age seventeen he had built his own oscilloscope. He studied at the Ecole Polytechnique in Paris, which is the elite school for any Frenchman interested in hard science. Sadoulet claims his academic career was nothing sterling—he graduated near the top of his year, but no better. However, one Frenchman told me, "All French physicists know of Bernard Sadoulet."

Sadoulet went to CERN to work on a bubble-chamber team for his thesis, and did not enjoy it. "Bubble chambers are much too serious for physicists to get close to," he told me. "I like to have an oscilloscope and to look at something; to have a soldering iron to build something." He wrote his thesis virtually without guidance, and afterward moved on to electronics and computers. In 1972, his temporary contract at CERN expired. At twenty-eight, he was too young for a permanent position, but that was the only factor not in his favor. CERN asked him to get lost for a couple of years and come back when he turned thirty. He went off to Berkeley to work with the Mark I experiment at SLAC. There, even while working on the experiment that discovered the J/psi, Sadoulet managed to become

*If calculated in American accounting instead of European, it would be roughly twice as much, as the Americans include labor in the cost and the Europeans do not.

a member of the "I Missed the J/Psi Club." The mistake of his career, he calls it. He was scheduled to be on shift the day of the discovery, but opted instead to stay at home and work on a future proposal.

When Sadoulet returned to CERN, he was considered the hottest prospect around. "That genius Sadoulet is back," said the rumor mill. Sadoulet was wooed by various experiments, especially those that were just cranking up and might last longer than the three years of most appointments. Eventually, he threw in with Rubbia, although he had no misconceptions about Rubbia's reputation, which was "very bad, extremely bad."

Sadoulet had known Rubbia back in the late sixties and had not greatly admired him even then. Sadoulet has a rather refined, intellectual sensibility, and he was amused by Rubbia's temper and his politicking. But as Sadoulet said later, "Carlo seduced me in a sense. He had really *the* project at CERN."

Once Sadoulet joined, and ICE looked like it would fly, he and Rubbia took to barnstorming. Whenever they heard of a group of physicists who might be looking for an experiment to join, one or both would visit and give their spiel. They told them about the marvelous technology they would use in the detector. They sold them the W and the Z, and the opportunity to work on the most exciting experiment of the decade, on the machine that would be at the edge of the universe, that would work at an energy never before reached. One physicist described it as the Rubbia roadshow: polished transparencies, lots of hyperbole, plenty of striding back and forth and gesticulating, and remarkable amounts of enthusiasm.

The response was not overwhelming. In Germany, at the Physikalisches Institut in Aachen, only four physicists were interested and only one at first, an old Rubbia disciple who had learned over the years how to work with him. "For me it was clear that it was the only experiment to be done, to find the W and the Z," he said. "If you feel an experiment is very good, you cannot hesitate and say, I do not like to work with the guy." In France, at the Centre pour l'Energie Nuclear Saclay near Paris, only two were sold, and those two were bubble-chamber physicists who knew that they were reaching the end of their careers with the antiquated technology of bubble chambers and that the time had come to develop new skills. As long as they had everything to learn from Rubbia, working for him might be almost pleasant.

In Britain, the response was equally ambivalent. They liked the physics, they did not want to work for the man. The British had a

different philosophy. "We are very conservative," said Peter Kalmus, the leader of the high energy physics group at Queen Mary
College in London, "not particularly brilliant, and as far as I know,
every experiment we've ever done has been right, but not particularly interesting." The British knew, all too well, about Rubbia.
Some of them had worked with him in the past; one swore he would
never do so again in the future, although eventually he did. Eric
Eisenhandler, a QMC physicist, recalled that in the early seventies
the British contingent at CERN had offices that faced Rubbia's conference room. "We used to watch their group meetings through the
window," Eisenhandler said. "Watch Carlo shouting at people and
shredding people. So we sort of knew him."

But they joined. They joined because the British government had
just shut down their only two national accelerators to save money
and they had to do something. They joined because they thought the
physics was worth it, because it was "the project at CERN." They
joined, as one physicist told his colleagues at the time, because "if
one has to do physics, one might as well make a big move." They
joined because, as Kalmus said, "perhaps if you have one brilliant,
impatient physicist and others more steady who will check to make
sure things are done thoroughly and not prematurely, then . . ."

As Rubbia was building his team, the CERN management was considering how they could ease their own fears. UA1 would have to
find the W and the Z or CERN would be more than just embarrassed
—they might never climb out of the public image hole into which it
would drop them. They were relying on the cash and goodwill of at
that time twelve European governments, and it was time they paid
it back with results. And, as Adams had demonstrated, they had no
delusions about the man on whom their future depended. "Rubbia
had done good experiments," Van Hove recounted. "None were
very good. And he left them very quickly when they were not giving
the beautiful things that he had expected when he started. It was,
of course, a very important fact for us to remember, that you could
not rely totally on Rubbia's critical judgment, nor on his own sense
of continuity."

The CERN management first set about moving Rubbia off his
other experiments. They knew that many of his problems had come
when he had spread himself too thin. By early 1978, he had split from
Fermilab and the ISR, and his Brookhaven experiment had been run
for years by an assistant professor. But now he had an experiment

at CERN known as NA4, for North Area Four, on the SPS muon beam. The experiment had been approved before the collider, but CERN wanted Rubbia off. "We moved him out," said Van Hove, "perhaps a little faster than he would have liked."

Rubbia did not go easily. He complained that the management treated him like a schoolboy, that they had set up conditions under which he could have his colliding-beam experiment, and that they had then told him he must leave NA4 and do it their way. Rubbia wanted to hand over the spokesmanship for the NA4 experiment to one of his disciples. He had lots of arguments. He said, "If you build something, you want to leave it in the hands of the best person you know." And: "For me to ensure the continuity of the program, I wanted to see younger people taking over." And: "I built that detector, put the group together. It was my idea, a good idea, and I was completely against giving the job of being spokesman to this guy who was a sheer political figure."

All these arguments sounded eminently reasonable. But the management knew what Rubbia had done in the past. They did not want him oscillating between the two experiments should NA4 somehow turn up an interesting result. (A young American physicist described the technique as he lived through it on one of Rubbia's earlier ISR experiments: "After the experiment was built, we were just beginning to take data. At that point he made the move to come back and take over and assume his normal leadership role. Even the junior people were tired of this. He comes in and plays around, and you feel like you're manipulated like a puppet, and when the time comes to get the results of your hard-earned work, someone else is going to come in and do something different. Well, you just tell him to go to hell.") CERN eventually forced Rubbia to leave the NA4 experiment in the hands of the man he called a sheer political figure, a senior German physicist, and cut Rubbia off from any future participation. His NA4 collaborators, as well, asked him to leave after tolerating a year of his leadership in absentia while he concentrated on pbarp.

Rubbia felt that the management had ulterior motives in getting him out of the muon business. Erwin Gabathuler, a British physicist from Daresbury, was division leader of experimental physics at CERN, as well as the spokesman for a huge experiment known as the Munich Collaboration, or NA2, which was upstream on the same muon line as Rubbia's NA4. Earlier, Rubbia and Gabathuler had argued about beam time. Rubbia had complained to the management that he was, in effect, getting a poor muon beam because the particles first passed through Gabathuler's NA2. The management decreed that for half of the time the physicists in control of the accelerator would focus the beam on Gabathuler's NA2 experiment, and the other half on Rubbia's NA4. When the beam was focused on one experiment, the other would get lousy beam. Having assured that NA2's data-collecting time had been effectively halved, Rubbia then turned around and ordered his physicists to take data all the time, whether they were getting good beam or not. Now, Rubbia was convinced that Gabathuler was getting revenge, trying to kill off his competition. "He was very concerned, very afraid," Rubbia told me. "The Munich Collaboration spent twice as much money as UA1." (And he added, "They haven't got a fucking thing out of it.")

When not fighting with Rubbia on NA4, the management was con-scripting some of the best physicists from around the lab and send-ing them off to work on UA1, to provide, in Van Hove's words, "the technical strength and soundness of judgment that is required." Van Hove was particularly worried about the enormous amount of com-puter software that would be needed to analyze the data from the massive experiment, so he had his research director ask Rudy Bock, a German software expert with twenty years' experience at the lab, to work on the computer programming for the detector. Bock had made plans to take a few years' sabbatical at SLAC, but the man-agement insisted that the experiment "needs people who can keep Carlo on his rails, but quietly." So he stayed. The management also asked an Italian physicist named Pier Giorgi Innocenti to join UA1. Innocenti had been running the Split Field Magnet experiment at the ISR, a huge detector that was used as a facility around which col-laborations added their own equipment. Innocenti was one of the most respected men at CERN, and Van Hove hoped he would serve as Rubbia's second in command. It wasn't to be the case. Innocenti stayed for six months and then left, because he said, "I didn't accept Carlo's way of managing, if you want." Still, Innocenti had brought with him a team of people from the SFM that included some of the best software and detector physicists at CERN, and when he left, they remained.

The final UA1 team included physicists from three institutions in France; three in England; and one each in the United States (Univer-sity of California at Riverside), Germany, Austria, Finland, and Italy, as well as people from CERN. Most of those physicists had been working on the ISR and Rubbia's various old experiments. Eventually there would be 135 names on the paper that would restore Rubbia's reputation.

In 1979, CERN engineers began to excavate a pit several hundred yards from the French border and six stories deep. This pit would become the home of the experiment known as Underground Area One, or UA1, which is how the physicists refer sometimes to the pit itself, sometimes to the detector, and sometimes to the collaboration of physicists who had joined up to work in it.

When completed, Rubbia's collider would take two focused beams of particles—one of protons, one of antiprotons—and accelerate them in opposite directions through the four miles of vacuum-filled pipes at a hairbreadth less than the speed of light. Each beam would actually consist of three bunches of particles, each less than a meter long and a millimeter across, and those bunches would pass through each other like ghosts 44,000 times a second. One time in every dozen or so passes, a proton would collide with an antiproton. With each collision, matter and antimatter would annihilate each other, disintegrating into pure energy, which would then congeal again into new particles. Those particles, in turn, would decay into nuclear debris that would fragment and shatter from the extraordinary strengths of the quantum forces, spinning off into the electronic eyes of UA1.

One time in every billion or so collisions, the right breed of quark, at just the right energy from the proton, would smack head on with the right breed of antiquark, at just the right energy from the antiproton—or at least so the theorists predicted—and create a W particle. It would exist in the heart of the detector for perhaps one billionth of a trillionth of a second before exploding into a shower of more enduring particles. The detector would have to provide the physicists with enough information on that infinitesimal explosion from that one out of a billion collision so that they could recognize the W by the remnants it had left behind.

"Detectors," Rubbia once said, with his usual hyperbole, "are really the way you express yourself. To say somehow what you have in your guts. In the case of painters, it's painting. In the case of sculptors, it's sculpture. In the case of experimental physicists, it's detectors. The detector is the image of the guy who designed it."

For UA1, Rubbia shot for a detector of an unprecedented scale. He had proposed such detectors before at CERN, but they had been rejected as being too ambitious by the various committees, which objected consciously or subconsciously to radical ideas. "The point," said Rubbia, "was that by pushing through the collider program, we were in no-man's-land. I took that opportunity to push through something that was in my guts for a long time. I would never be able to push it on its own merits. They were men of compromise. Give them a proposal, they cut half of it. You can't cut half of this detector, there'd be no detector left."

This one the management approved virtually without comment, without modification. It was Rubbia's dream: the ultimate detector. It would surround the collision point on all sides, as close to being hermetically sealed as possible, so that none of the tens or hundreds of particles spat out of a collision would be able to pass through without escaping notice. It would have computers on computers on computers. It would capture the image of the collisions in three-dimensional glory.

The philosophy of the detector evolved from the lessons of the failures at the Intersecting Storage Ring. The ISR had been the first and only proton-proton collider and, in part, the physicists had missed all the physics because they had had no experience with such colliders. Physics at the ISR, by decree of the CERN management as well as by the mistaken judgment of the physicists, was what Rubbia called keyhole physics: a hundred physicists all looking at a collision through a hundred different keyholes.

Imagine a sphere surrounding the point at which protons collide. At the ISR, the management allotted each experiment one small wedge of that sphere in which to erect their detectors, and to capture whatever particles came through that wedge. Each experiment got a slice, as though they were serving watermelons at a party and had to have enough for everybody. "I remember we had the angle between 8 milliradians and 120 milliradians," Rubbia said. "That was our domain. Now, Amaldi had the angle from 10 milliradians down to 1 milliradian. Somebody else had the angle between 100 milliradians and . . . etc. That was really the philosophy of the so-called big bosses at the time, some of whom are still hanging around, unfortunately."

The ISR had seven working interaction regions in which the beams collided, and these were expected to serve a community of several hundred physicists. To Rubbia, who believes physics should be a meritocracy, the protocol at the ISR was a function of the laboratory's multinational personality. "CERN itself was all divided into slices," said Rubbia. "It was the bureaucrats who were running the machine. And they said, no physicists can have the whole lot because that will be unfair to the others, so why don't we cut it out in small angles? It was a great discovery from the point of view of politics. It was a disaster from the point of view of science."

Rubbia immediately planned for a four-pi detector (in the language of geometry, the surface area of a sphere is equal to four pi times its radius squared) that would completely surround the collision point. It would do the physics of all the myriad experiments of the ISR at one time, and any particles that were created in proton-antiproton collisions would have to pass through its electronics. The concept had been originated at SLAC by Richter; all his colliding-beam detectors had been four-pi.

The actual detector would then be what Rubbia once described as "a series of boxes, each one doing what the previous one couldn't do." The innermost box was the central detector—the chamber that would take the collision of a proton and an antiproton and capture it as a three-dimensional image. But this central detector, or CD, as they called it, would mark only the passage of charged particles—and it would not provide enough information to positively identify even those. Neutral particles would be completely invisible to it. Outside of the CD were "boxes" that completed the task of identification. First, the electromagnetic calorimeters, or gondolas, which were cylindrical layers of detector made of a type of plastic, called scintillator, that emits light when a particle passes through it. The gondolas would identify particles that were affected by the electromagnetic force, such as electrons and photons, by measuring the amount of energy deposited when they hit the scintillator. Next came the hadron calorimeters, or C's, which were rectangular layers of iron plates interwoven with scintillators that would recognize hadrons, the particles that obey the strong force. Around all this came a layer of parallel wires enclosed in rectangular gas-filled chambers that would identify muons (heavier brothers of the electrons), which would have passed through both the calorimeters virtually untouched and unnoticed.

After the UA1 proposal was accepted, the building of the detector was apportioned out to the various European institutions. The central detector was the choicest assignment because it was the most technologically challenging. No one ever doubted that it would go to the CERN contingent of UA1, and be built under Rubbia's eyes, as indeed it was. Next were the gondolas. The physicists hoped to identify a W when it decayed into an electron and a neutrino (which the theory said it would do no more than a few percent of the time), and a Z when it decayed into two electrons. The gondolas would be the critical piece of apparatus for identifying the electrons, so whoever built them would have some measure of control over the analysis. The British argued for the gondolas, but Rubbia gave them to the French from Saclay as a reward for their early support of the collider project. The British settled for the hadron calorimeters. The French from Annecy took the bouchons, the plugs that fit in the ends of the cylinder-shaped detector. The Austrians, a small contingent, took the electronics that would be needed to read out the signals from the gondolas. The Germans, all four of them, took the muon chambers, which while they were conventional technology were almost as large as the rest of the detector.

The physicists—or the technicians and engineers of their various institutions—then designed every piece of equipment, every software program, every electronic device for the detector. Unlike the space program and the military, which ask industry to provide the product, the physicists, even with $20 million to play with, could not afford to leave the planning to anyone but themselves. "If it turns out that the magnet does not work, or the detector does not track," Rubbia said, "we cannot blame anybody but ourselves. You have to understand every part in your detector. You have to know what makes it run, how it works."

For three years, physicists worked as technicians and bureaucrats. The West Germans from Aachen led by thirty-five-year-old Karsten Eggert, spent the better part of the first year arguing with their government to come through with the nearly 5 million deutsch marks (roughly $2 million), the largest amount ever provided by the German government for a single experiment. They spent the next two years driving through Germany, covering maybe 25,000 kilometers, looking for an industry that could build the muon chambers. They looked at the high-tech industries and the space industries. "We were thinking," said Eggert, "that the industry that makes Spacelab should be able to do these comparably simple chambers." They found out that because they were not paying as much as the defense establishment, contractors would not put their best people on the muon chambers, but instead would assign temporary help, workers who weren't particularly motivated by contributing to the search for some esoteric piece of physics. After the chambers were built, the German team spent nearly half their time going back to the factories to clear up the mistakes.

At Saclay, outside Paris, the French learned that they had promised that the scintillator in their gondolas would turn out five times more light than was physically possible. They spent the next two years rethinking scintillator technology, and searching for a plastic that would give them enough light without costing more than the whole UA1 detector put together. When they found one that would do the trick, it turned out to have been purified by a German chemist in his spare time at home. So they talked the chemist's firm into developing the plastic as a research project, which meant that they could get the 12 tons they required. Then they found a shower-stall manufacturer in Belgium that would turn the German plastic into high-tech physics apparatus. They moved their computers and their sophisticated testing equipment into the shower factory to check the plastic as it rolled off the assembly lines. Not until 1980 were they confident that they were not cooking up a fiasco.

Meanwhile, the British found themselves spending those years running a clearinghouse for high-tech physics paraphernalia. Eric Eisenhandler, a Brooklyn-born physicist who since 1968 has been at Queen Mary College in London, described the work on the hadron calorimeters, the C's, as a cross between accounting and assembly-line work. "Seven thousand pieces of scintillator weighing a total of forty tons," Eisenhandler enumerated. "Eleven thousand wave shifter bars, 11,000 light guides all twisted and curved around in funny ways. Some rather high-quality photo tubes. I was involved with the tests of the phototubes, and then I was involved in the very exciting business of getting 1,200 (1,500 allowing for spares) photo-tubes ordered, specified, delivered, and tested. And we did test them quite thoroughly. It was boring. This lasted about two years. . . . You spend most of your working time not doing physics. You're learning whether this or that works, going through the hadron calorimeter looking for light leaks, going through the electronics looking for dud chips or bits of solder that somebody has dropped between two wires that have shorted something out. There's a lot of slum."

And while they worked, a group of CERN physicists created a computer system that would have to handle information as quickly as three or four of the world's fastest supercomputers combined. If the detector was the eye that would see the collisions, the electronic readout, the computers, and the software would be the brain that would interpret what they saw. If the brain could not work fast enough, the information from the eyes would be lost. Many of these people knew little physics, but they knew computers as well as or better than anyone you could find in Silicon Valley. As one physicist told me, these guys were so valuable that if they were to disappear, CERN might as well take the $20 million UA1 detector and sell it for scrap.

While the UA1 team worked, they sweated under the pressure. The collider would be ready by the summer of 1981 and the detector would have to be sitting in its pit by then, waiting for the collisions. They could not bring the product in six months or a year late and expect the machine to wait. And that was not the only pressure. For the first time in their lives, they were working on an experiment that was newsworthy, and the management of CERN was going out of their way to make sure that the world's press knew it was newsworthy. The management had their reputation on the line as much as the physicists did. It seemed that every week another journalist was in asking questions, or there was an article in the local *CERN Courier* or even more local *CERN Bulletin* showing a picture of Rubbia and detailing the progress on the machine. Two TV crews visited UA1 during this time. On top of all this they had the simple pressure of working fifteen hours a day, seven days a week, for maybe five or six years or even more, before they saw the payoff —if there was a payoff.

Every year the CERN theory department staged a Christmas play, a satirical play that poked fun at the politics and the personalities of the lab. Two years into the project the theorists sang "The Song of the Dejected Experimentalist." It might have become the theme song of the UA1 team: "What is physics coming to?" they sang, with dejected mutterings in the background. "It's a mess. I think I'll become an engineer . . . At least they get tea breaks."

The collaboration would meet monthly back at CERN to present status reports. They would cover the entire detector in a day, starting from the inside, the central detector, and working slowly out through the gondolas, the hadron calorimeter, the muon chambers, then out of the machine and down the beam lines of the collider. Once a month, a physicist would stand for fifteen or twenty minutes and tell his colleagues, who maybe barely knew him, if at all, what his team had done, or what they had not done.

Through it all, Rubbia was on top of everything. He would still be flying off to Harvard to teach courses, or giving a lecture in some distant country to raise money or recruit physicists. But he would control the detector to the best of his paranoid abilities. "Paranoid" is the word Rubbia himself uses. He would push his physicists to work on a timescale that they considered impossible. He would tell them he wanted some device in a weekend that they thought would need three months, and they would eventually get it to him in two weeks. They would never know quite how they were able to do it so fast. But Rubbia expected it, and it wasn't worthwhile giving him another reason to scream, since he found so many without their help.

When they did run into problems, Rubbia was the one, more often than not, who got them out. "He was very good at cutting out the crap that exists around things," recalled Alan Astbury, a cheerful British physicist who was co-spokesman of the experiment, "and pointing people and the experiment in a very straightforward, simple direction. If you could get him into the meeting. (Well, he was a hell of a lot more available then than he was later on; he was basically here every day.) That was when he really stood out; not just as a physicist, but almost as somebody seeing his way through a maze. He would listen to what somebody said, and then ask a few questions, get a bit excited—if he didn't think you were doing what he thought you ought to have done—and then say, 'Well, look, you know, here's what you do.' And it used to make a lot of sense."

When it was all over, they would say that perhaps the oddest aspect of this existence was that they worked in separate groups, each of which was as large as the largest collaborations in which they had ever previously taken part, but which was still only one small part of UA1. They worked on a piece of the apparatus that was about as large as an entire experiment used to be, and they concentrated on that single piece. They did it to the exclusion of most of the rest of their lives. In the past, they had understood everything that went on in their experiments. Now they had an understanding of only one facet. They prayed not only that their piece of the apparatus worked, but also that another group's didn't turn up as, in one physicist's words, "a pile of rubbish." They had no idea what was really happening with the other parts of the experiment. They only knew what people admitted in the collaboration meetings, and what they overheard in corridors.

Many of the physicists suspected that at least one critical piece of the detector wasn't going quite as well as it should be going, but they never knew for sure. They would notice what seemed to be a lot of hustling and commotion surrounding that one piece. A small group of physicists and technicians would be in the lab early in the morning when they arrived, and would still be there late at night when they left, appearing always to be in some kind of trouble. Indeed, this one group seemed to be in a crisis situation for five years. It was the group that was building the central detector.

Since the detector had to be built fast, since it would be huge, and since it would have to work first time out, its creators decided that it should use conventional technology. (Ex-SLAC director Wolfgang Panofsky once said that in physics, "the definition of conventional technology is that it worked once.") But one crucial piece would be a gamble: the central detector. The CD was more than the heart of the detector, it was the eye; the innermost box of all the boxes, the chamber that would bring the event to life.

The central detector was designed by Sadoulet and Rubbia with concepts borrowed from Burt Richter's experiments at SLAC. But both Rubbia and Sadoulet wanted to push the technology farther than Richter ever had. "Richter had persistently chosen techniques, at least for Mark I, which were a little outdated," Sadoulet told me. "But he wanted to be sure. That's not Carlo's style, and that's not really my style." They were also going to build it fast. Sadoulet and his crew would get three years to build their central detector. For the SLAC device, a team from Berkeley, headed by Dave Nygren, had taken more than twice that long.

The UA1 central detector would be an advanced version of the SLAC device. It would be larger, faster, and the three dimensional images it afforded would be more dramatic. The plan called for a gas-filled cylindrical chamber about six yards long and two and a half yards wide, with six thousand parallel "sense" wires crossing the volume. The collision of proton and antiproton in the vertex of the chamber would spew particles through the central detector, and these particles would rip electrons out of the gas as they went. The free electrons would then be caught in an electric field that flowed through the chamber, and would drift in that field over to the sense wires. In much the same way that a television scans electrons across a screen to produce an image, the electric field in the chamber would pull the electrons across to the sense wires. When the electrons hit the wire, they would deposit their charge, and from the patterns of those charges across the six thousand wires the computers of UA1 could put together an image, a reconstruction of the trajectories of the particles spewed from the original collision.

Having collaborated on the original conception of the CD, Rubbia and Sadoulet then diverged in their activities and preoccupations. Sadoulet designed the mechanics of the device and tried to keep Rubbia's participation to a minimum, since he believed that Rubbia's personal style was simply incompatible with building chambers. "Carlo is too fast," said Sadoulet. "He's not systematic enough. A chamber is not something that you put together and it works. It is something that is at the margin of feasibility. You have to work very carefully. And to a large extent, Carlo had not succeeded in his groups to build working chambers." Rubbia concentrated on the electronics, his forte, and designed the readout system, which takes the pulses of the central detector and turns them into signals that can be used by the computers or processors. The readout system used the only truly new technology in the apparatus: electronics that were faster than anything ever used before. It would allow the computers to pinpoint the position of the particle tracks in the detector to within a quarter of a millimeter on six thousand different wires, and to do it every four millionths of a second.

From the very beginning, Rubbia and Sadoulet clashed. Sadoulet considered Rubbia's management style that of "the Italian professor trying to check on his students," and he organized the central detector team in a very "non-Rubbiaesque way," trying to anticipate problems with Rubbia. "Very specifically," said Sadoulet, "we didn't want him to have any responsibility in terms of manpower. We wanted to have the control, to prevent priorities from being reset every half a day. So, for instance, we managed things in such a way that we cut Carlo off from his previous groups; we did not accept any technicians from his NA4 experiment."

For his part, Rubbia was always irked by Sadoulet's proud intellectual demeanor. The fact that Sadoulet had a slight stutter didn't help. As one British physicist put it, "Sadoulet used to take a lot of shit from Carlo. Carlo used to humiliate him in public. It wasn't nice. Bernard is a nice guy." An American, a young Rubbia disciple, said bluntly, "Carlo hates Bernard's ass. Sadoulet is from the Ecole Polytechnique, very elite, very theoretically oriented, and Carlo thinks it's a piece of shit."

Sadoulet himself was often amazed at how easy it was for him to set Rubbia off. "I can get Carlo mad in thirty seconds," he said. "But at times, when I wanted to be completely innocent and not do it, a single word would trigger an explosion." And the explosions happened frequently.

Rubbia and Sadoulet disagreed on a lot of points and, without any buffer, it was always the two of them that fought things out. They disagreed on the type of magnet to use for the detector. (The charged particles that come out of the collision will curve in the magnetic field as the path of a thrown baseball curves under the pull of gravity. From the amount of curvature can be calculated the speed and energy of the particle. The topology of the magnet will determine the shape of the field and where it is strongest and where weakest.) Rubbia won that one.* They argued about how best to distribute the six thousand wires in the central detector, and Sadoulet won that one. They argued about the electronics for reading out the data from the chambers. Rubbia considered the electronics his private domain, but Sadoulet argued that he could not design the mechanics without knowing what went into the electronics. Eventually Rubbia won by throwing Sadoulet out of the electronics group and telling him that if he didn't concentrate solely on the mechanics, it wouldn't get done. They argued about the deadline for the detector, which seemed unreasonable to Sadoulet, and they argued about Sadoulet's hours, because he refused to work more than ten or eleven a day (although sometimes it was seven days a week). Sadoulet wanted to have some kind of family life, and he insisted that he could work better in ten hours rested than in twenty under strain. Rubbia didn't agree, but Sadoulet won that one.

Sadoulet thought of quitting many times because of the strain, and because of Rubbia, but he never did. "I went through some very bad experiences," he said, "just because the interest of the physics was so high." He was also certain that Rubbia was an amateur psychologist and that he treated people, and himself, badly in order to get the work done. It may or may not have been efficient, but it seemed to be Rubbia's management style. "If he had had the impression at any time that I could leave and drop the thing in midair," said Sadoulet, "he would have behaved slightly differently."

The central detector was built by a half-dozen physicists and five times that many technicians. They worked as though physics was all that existed in their world—there were no weekends, no nights

---

*In roughly thirty hours of interviews, Rubbia mentioned Sadoulet's name only once, in reference to the argument over the magnet, on which he had come out on top.

to sleep. Several of them risked divorce. One physicist's wife had a nervous breakdown. One physicist came down with hepatitis when it was finally over. Sadoulet remembers that during that period he read a book on the Manhattan Project and was struck by the parallels between the two projects. "We had no time to have several designs," he said. "We had to have one design and get it through, whatever the costs. We had no way but just to succeed the first time."

Looking back on it, Eric Eisenhandler simply said, "It was unbelievable. Those people worked much harder than anybody else. They had to. And they delivered the goods."

5.

THE COMPETITION

"In physics there's no number two. Who will remember what UA2 has done? Nobody will remember UA2."   —Samuel C. C. Ting, Nobel Laureate

"The thing you learn from past experience," Rubbia has said over and over, "is don't come up with a story too slowly. Come with a bang, so that everybody knows the thing is completely understood the day it's published. If you linger too much, too long, on doubts, uncertainties, people get tired, and they say, Well, this is a muddy business."

Bowing deeply to this logic, the management of CERN decided that they needed a second major experiment on the collider, this one to be known as UA2, for Underground Area Two. Originally, they had thought that the price of two would be prohibitive: since the experiments must be put in pits from 20 to 50 meters deep at the level of the SPS itself, the expense is greater than just the millions of dollars worth of electronics. Then they decided that without a second experiment, discoveries made by the first might not be confirmed for years. And if they were not confirmed, they might not be believed. (Steinberger's "Is There a High-y Anomaly?" paper was then barely a year old.)

The collider was a monumental gamble, and they were shooting for a monumental payoff. They knew from experience that if UA1 should announce the discovery of the intermediate vector bosons, and then have to wait four or five years for another laboratory somewhere else in the world to announce their confirmation, the impact would be diminished considerably. And the Nobel Prize would be delayed. UA2 would provide the insurance that nothing coming out of the collider was muddy. It would provide the echo that would demonstrate that the first bang was real.

In fact, few people really believed that Rubbia's grandiose detector would be ready for the collider, and they wanted a complementary experiment, a simple one using conventional, tested technology —a cheap experiment. UA2 would save face should UA1 not be finished on time, or simply not work once it was finished. CERN wanted a second detector that could find the W and the Z at a third of the price of UA1. As late as 1979, Van Hove was saying in committee meetings that it would be important to have other experiments on the collider because "it would be three years before UA1 sorted out the data from its central detector." They wanted a fail-safe to protect the lab and Europe from embarrassment.

Only two teams of scientists proposed experiments for the second hall. The French physicist Pierre Darriulat and the Italian Luigi DiLella led one of the groups. They had both been active at the ISR and had submitted their proposal in January, but had been ignored until the second hall had been approved. The other group was led by Sam Ting.

Both DiLella and Darriulat were in some ways Rubbia protégés. DiLella had known Rubbia since college. He is a frank man with an aggressive manner, thick gray hair, and gold-rim glasses that tend to slide down his nose. He had been three years behind Rubbia at Pisa, and then caught up with him again at CERN after Rubbia returned from the States. Together they worked on their first experiments at the new Proton Synchrotron. DiLella was twenty-three, Rubbia twenty-six. Although DiLella was impressed by Rubbia's talents, he found the experience, in his words, depressing. "I thought he always took pleasure in finding that he knew more than you," DiLella explained glumly. "He kept asking questions about physics or mathematics, and he was very happy if he hit the question that you were unable to answer. He also would not let anyone have any initiative. For example, he says you are in charge of constructing a high-pressure gas Cherenkov counter, but then instead of just letting you work on that and eventually having a judgment when the project is more or less finalized, he keeps looking over your shoulder at the design you made, and he keeps yelling at you. For example, 'Why, you stupid . . . you've designed a flange with eight holes for screws,' he says. 'Why eight? Why not six? Six could be as good.' "

When the experiments ended, both physicists stayed on at CERN, but DiLella vowed not to work with Rubbia again. They remained friends, however.

Darriulat, on the other hand, is, as DiLella put it, "one of the few guys who remembers the time he was with Carlo as a very good time." Darriulat smokes a pipe and looks a little like a French naval officer—which he was during his military service. He has a thick, earnest face and long, graying sideburns. He was a nuclear physicist at the Ecole Polytechnique, switched to high-energy physics in 1966, and had his baptism with Rubbia at the old Proton Synchrotron. He credits Rubbia with his particle-physics education and says he found him always stimulating and full of ideas. He stayed with Rubbia through the first year at the ISR, then broke off to do his own experiment when Rubbia began commuting to the States. "Then," he told me, "we got all the nuisance of Carlo and none of the benefits." Darriulat was known as a conservative physicist.

Both DiLella and Darriulat had experiments at the ISR in 1975, when they and several other physicists proposed upgrading the ISR with superconducting magnets. They wrote an open letter to CERN claiming that this was less risky than the pbarp scheme. When the management informed them otherwise, they decided—with two French physicists, Marcel Banner and Jean-Marc Gaillard—to put in a proposal for a pbarp experiment. It was a much more modest scheme than Rubbia's. The difference is evident just from the written proposals: Rubbia's UA1 proposal ran to 155 pages; DiLella and Darriulat's UA2 proposal to 34 pages. Their detector was designed specifically to find the W and the Z by looking for the decays, as the theory predicted. "It is a specialized device, designed to catch the W and Z," was the way Rubbia described it. "UA1, on the other hand, is designed to be a kind of Land Rover which is used to travel on the moon, a general all-terrain device."

The other proposal for UA2 was submitted by Sam Ting, who at the time had an experiment at the ISR. He proposed a much larger, more extravagant version for the collider. He had designed an apparatus that would absorb all the particles produced by the collisions except the muons, and would then precisely measure the energy and momentum of those muons. Both the W and the Z would decay by emitting either electrons or muons. And while UA1 had chambers to detect muons, it was designed to concentrate on electrons.

Ting was the early favorite at CERN. Many thought that the various committees would be unable to turn down the allure of a Nobel Laureate on such a controversial project. There were even rumors that Rubbia was trying to torpedo Ting, because he was afraid the competition might be too stiff from a Nobel Laureate and that Ting might get in the way of Rubbia's own hoped-for Nobel. Rubbia denied these rumors. He said later that he thought Ting's proposal was more powerful than Darriulat and DiLella's, and that it would represent "more ferocious competition," but also that Ting had said all along that he would leave the W and Z for Rubbia. "From the beginning he said, that's your baby," Rubbia recalled. "I don't believe that I would have gotten any competition whatsoever from Sam. He looks like a very hard fighter, but he is a very honest fighter." And then Rubbia added, "Ting had already got his Nobel Prize. He had nothing to lose."

Ting said that he tried to design his detector to be complementary with UA1 and that Rubbia would enjoy the competition. But then Ting takes immense pride in the fact that, over his career, he has never published a wrong result and he has never published second. About his UA2 proposal, he said, "If I had not thought I would be first, I would not have done it."

In the second week of December, 1978, CERN held a shootout. Ting outlined his scheme, then answered questions. DiLella did the same for his. The CERN committee did something they rarely did, bringing in an outside physicist to judge the proposals. As one of the committee members put it, he was "somebody who had some degree of independence from the committee." He considered Ting's muon detector neither feasible nor comprehensive enough. "If it had been extremely simple," the committee member said, "going for broke, and cheap, it might have been approved. But it was stretching technology to the limit, and not doing very much either. It was voted against overwhelmingly."

Ting had proposed at the shootout that his drift chambers, which would measure the momentum and energy of the muons, would have an accuracy far better than that achieved by anything else at the time. Ting always held that in experimental physics, precision comes first. "He had to know the position of the wires to the size of something like a bacteria," DiLella explained later, "and there were many wires. And it turns out that even if you are very accurate in measuring the wires with the voltages off, as soon as you turn the voltages and the magnetic field on, there are electromagnetic forces that pull the wires. And then we asked him how would he correct for that effect. He said that he would personally measure continuously the tension on the wire and then he would predict by calculations of mechanics and electromagnetism where the wires were."

The CERN administration rejected Ting's scheme in favor of Darriulat and DiLella's. But after the shootout and before the official judgment was announced, the CERN theory division went to work. They performed their annual theory play, offering "a free interpretation of some ancient Geneva history—and some recent CERN history." They reenacted the shootout, in which the Gang of Four enters, led by the wicked Ching, Mandarin of Massachusetts (played by a visiting Chinese theorist). The Mandarin announced: "We have made successful pilgrimages before in Brookhaven and in the small tunnel in Meyrin. We have always finished our long marches in eighteen months. We check the tension in our shoelaces by hand every half hour."

Ting, who was not present, heard about the play from colleagues. Between the satire and the rejection that followed, he was incensed. According to CERN physicists, Ting tried—but failed—to use his powerful political connections in the People's Republic of China to have the physicist who played the Mandarin of Massachusetts deported back to China. ("Ting takes himself terribly seriously," Rubbia remarked about the episode.) He also wrote a letter to the chairman of the CERN Scientific Policy Committee, relating his poor history with the CERN management and ending:

> We are concerned about the wisdom of approving a proposal from a group, some of whose major proponents are known for their less than successful record at CERN, and refusing a proposal from a group who has accumulated a large amount of experience and success in that field of experimentation. . . . I am writing this to you in the hope that you will understand why compared to SLAC, DESY and BNL, so many at CERN have spent so much and achieved so little for so long a period.

Later, Ting blamed his rejection on perhaps not having the proper balance of Europeans in his collaboration. "To work at CERN," he told me, "you really need a strong European collaboration. I certainly disagree that the other experiment was better. I suspect political considerations played a large role at that time." Rubbia also believed that the judgment was politically motivated, but for different reasons. "I think the reason for the acceptance and rejection should be searched for in a nonphysical region," he said. "There was a strong French push essentially, and the man in charge, the director of research, was also French. He had a great sympathy for those people."

Since UA1 had a six-month headstart, neither Darriulat nor DiLella had the time to worry about whether they owed their future to politics. After the turmoil had settled at CERN, both collaborations dug in with their experiments.

The other physicists of the world ignored them, convinced that the CERN scheme was rolling merrily on its way to failure. Some prestigious physicists doubted that the Europeans—or anyone, for that matter—had the technical virtuosity needed to pull off such a complicated creation. Others claimed that even in the event that the collider did work, Rubbia's crew would never find the W or the Z.

These physicists likened the colliding of protons and antiprotons to throwing two garbage cans at each other. It would be difficult for UA1 or UA2 to recognize anything in the mess. Inside each proton and each antiproton are three quarks—three antiquarks in the antiprotons—and a swamp of gluons, the particles that transmit the strong force. In electron-positron machines, on the other hand, the collisions are between one single, pointlike electron and one single, pointlike positron. Richter liked to say that the ISR produced four to five times as many charmed particles as his electron-positron machine at SLAC, and yet, because of the mess, the garbage of proton-proton collisions, the discovery of charm, fell to him. "That's a signal to noise problem," said Richter. "Their signal was big, and their noise was much bigger." The same problem, he said, would beset Rubbia's experiment.

The W, said the critics, would simply be too difficult to spot, even if it did exist. The Z would be easier—according to the theory, anyway—but because the Z was heavier than the W, and because there was only one genus of Z as opposed to two of W's (one positive, one negative), they would have to make ten times as many collisions to find one Z. Physicists doubted whether the collider would have enough antiprotons channeled in a tight enough beam to find the W's, let alone the Z. If they produced only one tenth of the number predicted, that meant that instead of making five Z's a year, they would make only one every two years. And nobody would prove anything on one Z. Even European physicists who weren't involved with UA1 or UA2 said that the first Z would be found in CERN's proposed Large Electron Positron machine, LEP, which hadn't even been approved yet and on which construction wouldn't start until 1983.

In December, 1979, when Glashow, Weinberg, and Salam won their Nobel Prize for the theory that predicted the existence of the W and the Z, Glashow blithely predicted that only after the Europeans spent 1.3 billion Swiss francs on LEP might the world be presented with the existence of the Z. He also predicted that crazy things were going to happen, although maybe not as crazy as the Z not existing. At any rate, he said, "I will not give my medal back."

He made no mention of CERN's collider, or Rubbia, his colleague at Harvard.

Perhaps the general consensus at the time was best expressed by the CERN theorist who, having heard Rubbia lecture on the collider, suggested that his project was missing a name. "Why don't you call it Colliding Rings for Antiproton Proton?" he asked. "That is to say, C.R.A.P."

6.

EXHIBITION GAMES

"Why not? I mean, what the hell, I should have become a lawyer."
—Carlo Rubbia, on one of his administrative victories

CERN lies eight kilometers west of downtown Geneva on the Route de Meyrin, past an automated-radar speed cop, an indoor shopping mall, a bowling alley, the international airport, a Rolls-Royce dealership, and the offices of Martini & Rossi and Hewlett-Packard. Although the laboratory is technically in Meyrin, a suburban town that has all the architectural charm of a low-income housing project, it sits straddling the border with France, with vineyards behind it and a hill on which sheep often graze in front.

There is also a second lab, which is over the border in France and which was built primarily for the fixed-target physics of the Super Proton Synchrotron. To get to this French lab from the main lab, you drive through the customs and up a hill, then turn right onto a straight two-lane road that cuts across one third of the SPS. On a clear day, which may be infrequent in this corner of Europe, you can see Mont Blanc and the Alps, towering behind the city of Geneva.

Rubbia used to make the trip from Switzerland to France frequently. During the day he would come by the SPS control room whenever he had visitors, or VIP's. He would take them to see the sights: the computer controls for the SPS, the large screens on which they had programs that could teach neophytes how to control the beams; the television monitors high up on the walls that continuously reported the status of the accelerator.

At night, Rubbia would drive over looking for someone to talk to. He knew the young accelerator physicists on shift, and he would walk in and drop into a chair and start talking physics. Not necessarily about his experiment, but about general topics, whatever was on his mind. It would come out like water from a faucet, a gush of thoughts on physics and the future and the machines. Then, when he got tired, he would turn it off and go back to work at the main lab, or go home and sleep for four or five hours.

On July 9, 1981, Rubbia was in the SPS control room because the collider was about to run both protons and antiprotons for the first time ever. Rubbia was scared. People like Richter and Wolfgang Panofsky—the director of SLAC, a world authority on accelerators, and considered by physicists one of the smartest men in the world —had said that the machine might not work. Rubbia said later that as far as he was concerned, he was never an expert in anything, and those people were the real experts: "Their warning was sticking hard in my mind, and I was scared stiff the thing wouldn't work."

The entire process, begun as an idea that van der Meer considered "too far-fetched," had grown into an enormously complicated piece of machinery. It began with a device called an electrostatic generator, which tore protons free from hydrogen atoms with an electric field, and sent them off to be sped up in a linear accelerator, then in a booster accelerator, then into the Proton Synchrotron (PS), where they were accelerated up to 26 GeV. Then this proton beam of 10 trillion particles was smashed into a target, creating a shower of secondary particles out the other side that would include about 5 million antiprotons. These would be quickly separated out with magnetic fields, and shunted into a ring, called the Antiproton Accumulator (AA).

Using van der Meer's technique, the accumulator would cool down the motion of these antiprotons so that a bunch that once filled up cubic meters of space was compacted into a package the size and shape of a human hair. As one bunch was being cooled, every 2.4 seconds another bunch of protons would crash into the target, more antiprotons would be extracted, placed in the accumulator, stacked—in the local jargon—on the other antiprotons, and cooled. After some 60,000 pulses, 300 billion antiprotons would be circulating in the machine. (This sounds impressive, conceivably the largest collection of antimatter in the universe since the big bang. But it would weigh all of a trillionth of a gram—if it could be weighed—and if dropped in a glass of water and left to annihilate with the protons in the water, would produce enough energy to raise the temperature of the water in the glass by no more than one degree.)

After about twenty-four hours the "stack" would be full. More protons would then be created and accelerated through the series of machines, then sent in three bunches into the SPS. The accumulator would send its antiprotons back to the PS, where they would be accelerated up to 26 GeV (from 3.5 GeV in the AA). Then they too would be switched out of the PS like three railroad trains out of the station, and each shot down a beam line into the Super Proton Synchrotron to circulate in the opposite direction to that of the protons. Both beams were then accelerated to 270 GeV. Finally, they would collide.

Rubbia feared, as did everyone else, that each time the two beams collided they would give each other a minute kick out of their orbits, so that within a thousand orbits the kicks would build up and the beams would go crashing into the walls of the beam pipe and disappear. (All this at 44,000 turns a second, so that within a hundredth of a second the beams would be gone, and the physics with them.) It was known as the dreaded beam-beam interaction. "Only after the thing was approved," Rubbia said, "and it was approved so quickly that criticism could not build up—were we hit by all these problems. If we had had all these problems spelled out, nobody would ever have given us the money to do this." As one CERN accelerator physicist put it, "It frightened the pants off people."

On that July 9, they sat in the control room and waited for some sign that the beams had blown. And when it appeared that they hadn't, "Carlo was jumping up and down a bit," according to Lynn Evans, a machine physicist who was on the controls that night. Rubbia told me that he was more excited by the discovery that the collider would work than he was a year and a half later when he thought he had found a W particle. "There we were," he recalled, "and the beam was injected, and we looked at the damn thing and the beam was still there. And we looked again and it was still there, when they were supposed to kill each other in one hundredth of a second! That was the most exciting moment in my whole experience."

Rubbia left early the next morning for the year's big high-energy physics conference in Lisbon. With him, he took the proof he would need to demonstrate that protons and antiprotons had collided near Geneva.

The SPS ran protons and antiprotons sporadically throughout the summer, as the two collaborations continued work on their detectors. In November and December the machine ran a total of 140 hours of colliding beams. UA2 had their detector in and complete, as expected. UA1 had had some problems, mostly construction screw-ups. At one point they found that they couldn't fit the C's—the huge hadronic calorimeters, shaped like the letter C, made out of iron and each the size of a small school bus—into the pit where the detector was assembled because somebody had built the ceiling of the building 80 centimeters lower than specified. The crane that was supposed to rotate them and lower them into the pit was now also 80 centimeters too low. Since they couldn't raise the building, they brought in a mason and had him chip the excess off the floor of the pit. Once they had the C's in the pit, they fitted the gondolas—the electromagnetic calorimeters—inside, and the central detector inside the gondolas. But when they tried to close it all together, that didn't work either. Somebody had built the gondolas 4 or 5 centimeters larger than specs, and they had to add a wedge of iron to the top and bottom of the C-shaped magnets so that they could lock the thing up like a chastity belt around the central detector.

As for the central detector, they had finished the chambers but they had not finished the electronics needed to connect them to the outside world. The electronics were Rubbia's responsibility, but, said Sadoulet, Rubbia had not pushed his own people as hard as he had others at UA1. Consequently they were behind schedule. The management, which had been having regular meetings with the two collaborations, questioned why they were installing the CD without electronics. Rubbia immediately blamed the CERN technicians who had been assigned to help build the electronics. He then drafted many of Sadoulet's people to work alongside his own. The main draftees were Michael Rijssenbeek, a Dutchman, and Mario Calvetti, a talented young Italian physicist who had been Sadoulet's second in command.

Calvetti, while being remarkably capable, was also reputed to be perhaps the most severe workaholic in an experiment full of workaholics. (The quintessential Calvetti story was that when his wife was nine months pregnant with their first child, she called him up at work and said, "Mario, I think it's time. You have to come home." Calvetti said, "Let me finish one thing here." His wife said, "Okay. Meet me at the hospital." Five hours later, Calvetti slapped his palm to his forehead, remembering what his wife had said. He called the hospital, and was told he had a new son. When his wife was pregnant with their second child, he was present throughout.)

With the help of Calvetti and Rijssenbeek, Rubbia had the electronics for the CD tested and into the detector in time for the November run, but not without problems. They could tell that the detector was working, but they just couldn't see anything with it. This situation was described emphatically to me as a "madhouse" by Daryl Dibitonto, a young Harvard physicist who had been working from the beginning with Rubbia on electronics. "Nothing worked," he told me. "It seemed to be going from one disaster to the next. Eventually the CD ran, and then little by little the goddamned system started to fall apart. No joke."

The CD gang went back to working twenty-four hours a day on the electronics and the readouts. They got help from the industries that had built the parts that were not working. And they got help from every institution in UA1, from each of which Rubbia drafted a couple of physicists, tossing them into the pot for the central detector.

By April of 1982, Rubbia's growing power at CERN had become evident. He was no longer just a higher-echelon staff physicist. He had become suddenly some kind of anomaly, a rogue prince in a stiffly controlled monarchy; a man who had something that everybody needed, which somehow set him above everyone else. Ugo Amaldi, a senior CERN physicist, once explained the theoretical power structure of CERN this way: "According to our constitution, which exists but which nobody reads, everything is in the hands of the director general. He can do whatever he wants." But in the spring of 1982, it began to appear that CERN had bet so much on Rubbia that he might have an influence even greater than the director general's.

The SPS was due to switch back into its colliding-beam mode on April 26. UA2 was in and ready. The huge UA1 detector had been rolled into position on the beam line; the central detector was mechanically complete, pristine, ready to go. All they had left to do was to clean the vacuum tube that ran through the center of the detector. This meant heating it so as to dislodge stray oxygen atoms that might be clinging to the sides, which would ruin the vacuum in the beam pipe. They had to keep the central detector cool throughout this process by blowing air through two perforated hoses running along the vacuum pipe, which should have been a trivially easy task.

A UA1 technician first noticed that something had gone seriously wrong when he saw a cloud of red dust rising majestically from the interior of the CD. Much of the rest of the collaboration learned about it the next morning when they heard Rubbia screaming that disaster had struck. "It was clear how excited Carlo was," recalled Alan Astbury, the UA1 co-spokesman with Rubbia. "He was going around the entrance hall of the main building saying we were all going to have a three-year delay."

By then they knew that the air used to cool the central detector had not been pure, not even close. At the time, the SPS staff insisted that a pressure surge in the pipes had dislodged years of accumulated dirt. However, the UA1 physicists were of the opinion that somebody had screwed up. Astbury said, "I think they had set the thing up and gone away for lunch, and in the meantime somebody changed the compressed air supply." One way or the other, instead of cooling the detector, the pump had blown a cloud of dust and rust over its wires and electronics circuits.

The UA1 crew took a couple of days to disconnect the CD from the electrical and gas connections in the detector. They lifted it out with a crane and laid it open to appraise the damage. It was covered with dust—like the inside of a vacuum cleaner after the bag has broken. At first they feared that it might be wet dust, in which case they would have been in far greater trouble; but it was dry. They didn't tell the management that, however.

Rubbia immediately began lobbying to postpone the run. The last thing he wanted was for UA2 to be functioning while UA1 was sidelined. There was no way that UA1 was going to make it on time, he said, and it was not UA1's fault but that of the accelerator people running the SPS. Now the UA1 physicists would have to clean it up themselves, and they would have to do it delicately and thoroughly.

To get his postponement, Rubbia employed a repertoire that was variously described as being out of Italian opera and *Commedia dell'arte*. Rubbia argued vehemently that it would be a waste of resources to run the collider for a couple of months with only UA2's detector gathering data when UA1 would be ready by that time. He also said that while he could clean up the central detector, he certainly wasn't going to take responsibility for any consequences that might result from rushing the job. It was a consummate use of political scare tactics. "We were trying to terrify people," said Calvetti. "In the sense that if something were to happen, the responsibility would be theirs and not ours. Nobody in the CERN management was ready to take that responsibility."

Behind the politicking was Rubbia's fear of UA2 and his own reputation. The second experiment had been approved in the first place because CERN feared that UA1 would not get its act together in time, if at all. "Remember," Calvetti said, "if we didn't discover the W, it would have been done by UA2. That was something that was really terrifying Carlo at that time. He talked about it continually."

Darriulat of UA2 had written a memo to the management saying, "We shall make fools of ourselves if we further delay the next antiproton run." UA2 was ready to do physics, they wanted to get it in, and they expected Rubbia to help. And although in public Rubbia had said he too wanted an early run, in the meetings at CERN he and Astbury did nothing to help UA2. Once Rubbia had assured himself of a postponement, he was more than happy to wait as long as possible. "Carlo and I were a bit unpopular," Astbury recalled, "but it seemed to us that we would be in much better shape by the end of the year. There were all kinds of letters in *Nature* saying CERN was stupid, they have this collider, they have the edge on America, they didn't run it and all the rest of it . . . but it seemed that the best thing for UA1 was not to run until September."

As it turned out, CERN postponed the run completely until October. UA2 lost. (Sheldon Glashow put it quite concisely in his colorful way: "When some fool turned a knob and virtually destroyed UA1, or at least filled it with rat turds in some unpleasant fashion, the resolution to that difficulty was to delay the experiment, not to run UA2 with UA1 turned off.")

One week after the accident, *The New York Times* ran an article about it headlined "A Breakdown at CERN." In May, *Science,* the scientific community's journal of record, ran a three-page article with a headline that proclaimed "The 6-month delay due to an accident will give physicists a chance to fine-tune their detectors, but the whole affair is quite embarrassing." It quoted Rubbia saying sheepishly, at the annual APS meeting in Washington, that the accident was "one of those facts of life for an experimentalist."

If, in the aftermath, Rubbia seemed unruffled by the calamity, it was for a good reason. The actual severity of the accident was known only to a small circle of UA1 physicists, as was the reason why Rubbia was ready to wait until September, why he wasn't anxious to get to the W as soon as possible.

The accident was simply not as severe as Rubbia had made it

out to be. One UA1 technician said bluntly: "I went and watched this clean-up procedure, at which I laughed daily. They took paintbrushes and they took a cover off a preamplifier box and brushed off all the connectors, and they took the cover off the next preamplifier box and brushed it off." Hans Hoffmann, Rubbia's right-hand man on technical matters, said, "To be honest, there never was a real accident. A lot of dust got blown all over the CD so it had to be cleaned, we had to take it apart and clean it. Okay, we could have done that even with a short delay of the run." Sadoulet too admitted that it could have been cleaned in a week, but he preferred not to take any chances. He also, as he said, "let Carlo handle the politics of it." But even if the accident had not happened, he was convinced that Rubbia would have had to postpone the run anyway.

What Rubbia neglected to mention either to the CERN management or to UA2 was that, as everybody had feared, UA1 had not been ready to run. The electronics for the central detector were not ready. As Sadoulet said, "Carlo was responsible for the electronics, but he did not put the pressure on himself that he put on others."

Missing from the central detector was the control for the extraordinarily high voltages that course through its wires. This control would ensure that their remarkable piece of machinery did not torch itself. "The beauty of the CD," Calvetti explained to me, "is that it is simple, because there are only wires inside. There are thirty kilovolts in that chamber, but once you have drawn things in such a way that you don't have sparks, it will stay there forever—as though it was soldered in iron. If you don't make mistakes, you don't break it. So it's very important to have a sophisticated control system that is able to prevent mistakes."

They had built a sophisticated control system for the 1981 run, and it had fallen apart. Daryl Dibitonto built it under Rubbia's supervision, and he explained to me that Rubbia had wanted to use the latest technological components for the CD, but that they had neglected to read the fine print. "The problem with working with high voltage," Dibitonto said, "is you have to find yourself a device that electronically isolates the low voltage control, which is typically five volts, from the high voltage, which is running, in this case, at two thousand, maybe three thousand volts. So we needed some type of isolation devise. Carlo likes new things, new gadgets, and he found this technical flier showing this new chip. The blurb was saying, 'It's fantastic: six thousand volts, even as high as seven thousand volts, isolation between the low voltage and high voltage.

Complete electronic control, blah blah blah.' So Carlo said, 'You have to order these things,' and we went crazy and ordered about two hundred and fifty of them. To make a long story short, it turned out that the technical specs were not clear, and this famous seven-thousand-volt difference was only for Alternating Current. That means the peak voltage fluctuates up for two or three milli-seconds, then it comes back down. But we don't run AC. We run DC at three thousand volts steady, and little by little these goddamn chips started to break down."

Over the winter, they cannibalized the old system and built a new one. "In '82," Dibitonto continued, "we had to completely re-build the whole goddamned thing. Carlo wanted to put in micro-processors in the high voltage, and that system took a tremendous amount of fast rebuilding because the beam was coming back in April. Only three months to build that thing. We worked like dogs."

Without the accident, Dibitonto said, they would have finished because they had to finish. But it would have been, in one of Rubbia's favorite phrases, a bricollage job: something thrown to-gether out of spare parts and whatever is at hand. "The accident gave us another breather, which was just what we needed to get that system going," said Dibitonto. "We had tried to make a bricollage system, and it would have worked. I don't deny it. It had to have worked. But I'm sure it would have fallen apart after a week or so of running."

By the time this affair was over and the collider run was re-scheduled for October, the UA1 physicists were more than ever in awe of what Rubbia could do at CERN. It was not the kind of physics power structure they had lived with in the past. "We had to change the entire SPS schedule," said Kate Morgan, a UA1 techni-cian, "and we needed it not to be our fault. We needed it to be the SPS's fault. We needed not to have to say we're not ready, because at that point UA2 might have gotten the jump on UA1. Couldn't do that. In some way it was fun. You knew at that point you were tempting the gods, and you had to be amazed at Carlo's power. We were all enthralled. How could he do such a thing? He could change the run?"

Two months later UA2 took revenge.

Both collaborations had begun playing with their data early in 1982, looking through the collisions they had created for anything that might show signs of being interesting. Specifically, they had begun to look for a phenomenon known as jets.

Jets were the prize of the exhibition season. Jets were supposed to be the only visible signs of quarks or gluons when they were torn free of their protons and antiprotons in the collisions. According to theory, quarks cannot exist as single entities, but must always come, at least, in pairs. The strong force between two quarks increases in strength the farther those quarks are separated from each other. This is rather like the force that acts on the ends of a spring, which increases as they are pulled farther and farther apart. If a quark is hit head on by an incoming particle, the laws of the strong force say that it cannot simply be knocked free from its brother quark in the recoil. Instead, it will spark into a shower of new particles. Rather than seeing a single isolated quark shooting from such a collision, physicists should see a narrow cone of high-speed particles—a jet.

This counter-intuitive effect is at the heart of the theory known as quantum chromodynamics (QCD), which mathematically explains these vagaries of the strong force. It is not a well-understood theory, especially by experimentalists: Rubbia once tried to explain it to me, in what seemed a hopelessly convoluted way. When I asked why the quarks behaved the way they do, he replied, "Nobody has a fucking idea why. That's the way it goes. Golden rule number one: never ask those questions."

Jets had been spotted in electron-positron collisions at SLAC, and they had finally persuaded physicists of the existence of quarks. The jets may have been seen in hadron physics, hadrons being particles like protons that are made of quarks. Several experiments had claimed the discovery, but the rest of the world had not been thrilled by their evidence. Now, physicists wanted conclusive evidence of the production of jets in hadron-hadron collisions, and the head of the CERN theory department had very positively stated that they should be evident at the energy levels of their collider.

At CERN, throughout the winter and spring of 1982, a group of UA1 physicists had combed through their data for signs of jets in the central detector and had come up empty. At Saclay, three French UA1 physicists had looked for jets in the energy deposits in the electromagnetic calorimeters and had also come up with a murky picture.

The jet work at Saclay was headed by Daniel Denegri, forty-two, a Yugoslavian by birth. Denegri was convinced that he had found evidence for jets, but had failed to convince the physicists of UA1, who complained that he seemed to think in a circle instead of a straight line, and that when they listened to his lectures, they simply lost track of what he was saying. "If you're French," suggested one American physicist on UA1, "maybe you can understand perfectly."

In June, 1982, the UA1 collaboration met in Vienna, and Denegri explained his system for demonstrating the existence of jets. Nobody understood him, or believed him. "I was so disgusted with the way people took my presentation," Denegri recalled, "that I took my wife and two kids, and I went by car to Yugoslavia, and I said I don't care anymore. You don't want to listen to me so . . . go fuck yourself."

Rubbia, meanwhile, had become so convinced that the jets would not be found that he said as much in lectures he gave at CERN. According to Astbury, Rubbia once even went so far as to say that he questioned whether or not they would ever find the W, if the jet picture was so murky.

The major pbarp conference of the summer was in Paris. Several days before it began, Rubbia heard rumors that his competition might not be having quite so much trouble finding jets as UA1. One of Rubbia's lieutenants called Denegri in Yugoslavia and asked him to come back and present his results in Paris. Denegri refused, claiming it would be too hard to make plane connections in time.

In Paris, Allan Clark, an Australian physicist from UA2, talked about jets and presented beautiful, clear, unambiguous evidence. The speaker following him was from UA1. He made a presentation of very ambiguous data, the maybe-yes-maybe-no of jets. The rest of the collaboration went into shock. It was just what they had all feared, and what the cynics at CERN had been whispering the previous fall: While UA1 physicists jerked around with their huge, sophisticated, and complicated apparatus, UA2 had snuck in and done the job. (And there was no reason to expect it would be any different with them the next fall.) UA2 had actually had the jets all spring. They had heard Rubbia mistakenly predicting that jets would not be found, and they had smiled and kept their mouths shut.

In Paris, the UA1 physicists were struck immediately by the fact that all of UA2's jet events came from collisions in which a massive amount of energy had been released. (Protons and antiprotons are each composed of three quarks; the energy of the particles is distributed unevenly among those quarks.) The UA2 jets seemed to come from collisions in which the quarks in the impact must have carried the majority of the energy of the proton and the antiproton.

The UA1 physicists had never looked at their highest energy events, which had lain untouched in their data since December. They even remembered that a few weeks before, Rubbia had asked what was the highest energy event they'd taken so far, and whether anybody had looked at it. And they had said no. But nobody followed it up, and now UA2 had embarrassed them.

Rubbia called CERN immediately after the UA2 talk and told the UA1 physicists to pull the computer tapes of their collisions and take a look at the ones with the highest energy. Within ten hours of Rubbia's phone call, the UA1 physicists back at CERN were looking at jets.

Rubbia had the new jet pictures in his hand by the last day of the conference. But it was too late to do anything except save face. "It's not the same to come back and say a day later, 'Yeah, we've got jets too,'" said Astbury. "That battle was over. Why try to scramble anymore when UA2 had beaten us?" Instead, Rubbia retrenched. He told the assembly that the results proved that the collider was working beautifully, and that the jets were there just as predicted. He said UA2 had done a great job, as had UA1. "He didn't make an issue of it," said one UA1 physicist. "He sort of carried it off."

When they returned to CERN, however, Rubbia started to yell. Why the hell were their jets not there? Most of the blame fell on Denegri, and for the next year Rubbia would tell Denegri that he didn't trust him because he had failed him on the jets.

"Yes, UA2 found the jets first," a British physicist on UA1 told me. "Carlo's never forgiven them for that. He's never forgiven us."

7.

THE NOBEL RUNS

"I will give you my priority: to be first and to be correct is the most important. To be first but to be wrong I think is not as good." —Sam Ting

One summer night in 1982, about a month before the Super Proton Synchrotron was to start colliding beams once again, Carlo Rubbia was pacing up and down the balcony that ran along the upper floor of offices in the huge hanger-like structure that was buildings 20 and 21 in CERN's address system. The south side of the building was 20, where Rubbia's office was at the head of the stairs; the other side was 21. In between was a basketball-court-sized hangar, in which various huge pieces of equipment were always being built or torn apart or rebuilt or tested.

Rubbia had a newspaper folded in his hand, and he was waving it up and down and raving at Alan Astbury. The paper was a French daily, and the article that had upset him was about the coming run. It described the two experiments, and it did so, not surprisingly, from a Francophile point of view. It portrayed UA2 as the French hope, with Darriulat as its spokesman. UA1 was the juggernaut of CERN and Italy (the article seemed to ignore the French contingent of UA1). It portrayed UA2 as a small experiment, clever and cute. UA2 was David. UA1 was big and ugly and expensive, and it wasn't the kind of experiment you would build if you were going to do the physics of the decade. UA1 was Goliath. And it was pretty clear which experiment this French paper expected to succeed. At least that's how Rubbia saw it, and he was angry and seemed even a little frightened. The David and Goliath story was one of Rubbia's favorite metaphors; but hitherto he had always assumed the role of David, and now he was being portrayed as Goliath.

"What are we going to do?" Rubbia yelled. "What's it going to do to the group when they see this sort of crap? What if they're right?"

74

There was more to this than mere chauvinism on the part of one Parisian reporter. You could hear similar thoughts expressed any day in conversations in the CERN corridors or at the cantina where a good number of the world's physicists could be found at any time of the day or night. UA2 had good physicists—the jets had proved that. They did not have 130-odd personalities all trying to do everything at once. They worked well together. Sure, UA1 had their supposedly beautiful apparatus and their state-of-the-art central detector. But they wouldn't be able to figure out what it was telling them until years after the UA2 group, with their simple, uninspired, but sensible detector, had announced that they had discovered the W.

Even at UA1—even after the run had started—the physicists were saying that they shouldn't get too excited, because they might have nothing exciting to deliver. Even if the W were to show up smack in the middle of the detector, as they all hoped, they wondered if they'd recognize it.

Once it was created, if it was created, a W would exist for a billionth of a trillionth of a second and then—because it was too heavy to exist naturally—it would blow apart into various combinations of subatomic debris. The characteristic decay of the W—or the "signature," as the physicists would call it—that the UA1 physicists hoped to spot was a subtle one. They hoped to identify it by recognizing its electron. But there were a lot of other elementary particles that might look like electrons in the detector if the central detector wasn't working perfectly. And it wasn't, not yet at least.

A couple of hundred yards from Rubbia's office, through tho maze of buildings and barracks that make up CERN, the UA2 physicists were themselves beginning to believe that they could sneak in ahead of UA1 and grab the physics. But they had been thrown off stride when the spring run was delayed. They never doubted that the collider would run, but unlike Rubbia, they wondered how well it would run. For the spring run that never happened they had set up their detector to do bread-and-butter physics, the kind that ignored the possibility of finding the W's or Z's.

The UA2 dence was built something like an orange, with twenty-four slices of detectors looking in on the vertex where the particles collided. For the spring run they had removed about a quarter of the detector, and had installed a spectrometer that would let them do bread-and-butter physics—investigating the types of particles coming out of collisions, and their energy and direction. But the loss of that quarter of the detector meant that one out of every four W's created would vanish without a trace.

When the spring run was canceled, the UA2 physicists fought about whether to keep that wedge open or not. DiLella wanted to beat out Rubbia; he wanted to close that wedge and find the W. Darriulat, who was the spokesman, was more interested in just doing the best possible physics they could. They held a vote, and the result was that they left the wedge open.

The run began in early October, 1982. The machine was running well. Since the machine physicists preferred to work on repairs and timing during the day, they would not get the beams in until nine or ten at night. The beams would collide then until morning. Those physicists who had crucial jobs—who for instance had built sensitive portions of the machine, like the central detector—had to sit through the night taking data and babysitting the apparatus, and then catch whatever sleep they could during the day while their families lived normal lives, or tried to.

Rubbia rigidly controlled every aspect of the experiment. If the collider continued to run smoothly, they might see at most five to ten W's by the end of the year. They could not afford to miss any. As soon as the beams were in the machine, Rubbia wanted the detector cranked up and taking data. He did not want to see the physicists spending precious minutes dithering around getting it going. He wanted to make sure that by the end of the year they had at least had the chance to look at just about every collision that the SPS provided—all two billion of them.

Of course, even if their electronics could work fast enough to save every one of those two billion collisions on computer tape—which it couldn't—they would physically never be able to look at them all. And they would spend so much money on tape that they'd bankrupt the experiment. But only one out of every thousand events would be even remotely interesting, and the rest would be junk. The British at Rutherford Labs had designed and built a "trigger" system that would fire only on that one interesting event, and throw out the other 999. This trigger had to work extraordinarily fast, since the collisions happened at the rate of several thousand a second. It was not a computer, but an intricate system of super-fast electronic processors that examined all of the electrical pulses from the particles splattered into the calorimeters by the collision. If it registered a large splash of energy in a limited area of the calorimeters, for instance, it would interpret that as a potential electron, and tell the computers to save all the information from that event and then go onto the next one.

Rubbia had his physicists set up the detector and the computers so that within a day of a collision's occurring, they could scan the computerized reconstruction of that event on a machine known as a Megatek. These devices looked a little like oversized computer terminals, and worked like three-dimensional video games. The physicists could sit down in front of the screens and examine the computer reconstructions of the tracks from the collisions—and all the pertinent information that went with those tracks—as though they had taken snapshots and were showing them to their relatives after a vacation.

Early in the run they found what, at first glance, appeared to be a W, and Rubbia became very excited. It seemed to have a high-energy electron on one side and nothing on the other side, a possible indication that a neutrino had been emitted. Rubbia was leaving for the States shortly after it was found. He threw it in his bag, figuring he'd flash it to his colleagues at Harvard. But when he arrived in Cambridge, he got a call from Astbury telling him to keep quiet about the event. They didn't think it was a W. When they had studied it in detail, it appeared to have problems. As well as the electron they were looking for, there were a couple of small jets in the event, and those jets were shooting out in such a way that when they added up their energy it exactly balanced the energy in the electron. They suspected that whatever had created the two jets had decayed, ejecting the electron one way and the jets the other. They interpreted this to mean that perhaps they were not seeing the creation of a W, but instead the creation of two quarks, one fragmenting into two jets and the other into a jet that perfectly faked an electron. Physicists refer to such events as "background"—old physics that simply gets in the way when they're trying to find new physics.

When Rubbia returned to CERN, the collaboration met to discuss the event. Rubbia pushed it as a W, but a young British physicist named Martyn Corden refused to let him do it. "Somehow Carlo had gotten the impression," Astbury recalled, "that there were some people in UA1, Christ knows why, who were trying to sabotage it. He got himself into a peculiar state of mind." Corden had a reputation as a cautious physicist, and he refused to let Rubbia claim the event as a W. "Carlo was one of the first people to appreciate Martyn Corden's abilities," Astbury said, "but at that point in time he certainly didn't appreciate Martyn's, if you like, frank honesty in front of a hundred people of UA1."

Corden recalled later with a laugh how at that meeting Rubbia called him a "W killer."

After the meeting, the situation deteriorated further. Nothing even vaguely resembling a W showed up in the data. Everyone was nervous. A couple of the UA1 computer jocks, while working late night shifts, would sometimes patch into the UA2 computer programs to scan their files for any possible signs that they had W's.

And the CERN community was waiting; physicists kept asking if they had seen anything yet. If the W's did exist, if the standard model was correct, they should have seen at least one in the data they had collected.

Dave Cline, who had given up pushing pbarp in the States and was spending the fall of 1982 at UA1 as a visiting scientist, explained the feeling: "Everybody was saying the W must exist, it should exist, how can it not exist? But that doesn't mean a damn to you when you're trying to figure out whether these things are really W's or something else."

Meanwhile, Rubbia's fear turned to pessimism. He was desperate. He talked about publishing a paper claiming that the electroweak theory, for which Glashow, Weinberg, and Salam had already received their Nobel prizes, must be wrong. He told Alan Astbury one Tuesday in early November that if they didn't see anything by the following Sunday they would write this paper. But nobody ever won a Nobel Prize for proving that something didn't exist or by showing that somebody else was wrong. And if anyone knew that, it was Rubbia.

That Thursday they found the first good W candidate. It was beautiful, an incredible collision that had ejected particles all over the detector. Almost certainly it contained an electron with nearly half the energy of a W particle, just as the theorists had predicted. It had only one problem. One of the vagaries of the electroweak interactions was that W's were supposed to be created spinning like tops in the direction of the incoming protons—a circumstance caused by a characteristic known as helicity. And this one was not. The odds were five to one against its being a W. If the Weinberg–Salam model was correct, that was.

Rubbia argued that maybe W's had the wrong helicity. It was so damn close to being a W that it was almost inconceivable to him that it might not be one. He told Astbury that they should publish on that one event, and claim the discovery immediately. Rubbia argued that this case was no different than that of the Omega Minus, which was the classic one-event discovery. In the case of the Omega Minus, the theorists had predicted its existence down to the exact mass, and both a Brookhaven group and a CERN group had then raced off to find it. When the Americans found one event that fit the predicted characteristics and had the perfect mass—the theory had said 1.685 GeV, and the candidate weighed in at 1.686 plus or minus .012—they had published.

Astbury and his colleagues convinced Rubbia that this was a little different, that UA1 simply did not have enough confidence in their single event to make that kind of claim. But Rubbia leaked the story to the physics community (among other things, he gave a copy of the event to a SLAC physicist who had been visiting at UA1 so that he could take it back to California) to establish his own impetuous version of a priority claim.

Rubbia also took the event to the States with him. "We had a clear W signal in November," he told me later. "I went to Austin, Texas, in the first week of November, and I had with me in my suitcase a number of events which were certainly W events, beyond the shadow of a doubt. I showed them to a few very good people like Salam and Weinberg, but that was really absolute top secret." He also showed them to Glashow at Harvard, who predicted at a meeting of the National Association of Science Writers in Cambridge a few days later that the W would be discovered in CERN by his birthday in early December.

Rubbia took the event to Washington, and, with Cline and four other American collaborators, stopped in at the Department of Energy to flash it to the bureaucrats who dole out the funding for high-energy physics. Rubbia told them that they were the first to see the event, even before the director of CERN (this part was true). "We decided," Cline said, "that it was time to get more American involvement in the experiment. So it was time to go to Washington, and show them what we had."

The DOE people asked Rubbia and his gang to sign the computer-generated picture of the event Rubbia believed was a W, and then Bernie Hildebrand, one of the senior DOE men, mounted it on his wall. Later, Hildebrand heard second or third hand that Burt Richter had seen this event and had stated that he didn't believe it was a W. Hildebrand added to the signed picture a note saying, "Carlo, this is not a W," and signed it "Burt."

The run ended on December 7. Until then, the UA1 physicists worked their insane hours and nursed the machine and the central detector and tried to figure out how to prove that what they had were W's. By the time the collider shut down for the year, UA1 had maybe five events that might be W's, although they could not prove them. To make sure their physicists left the offices for the Christmas vacation and maybe even relaxed, the management not only turned off the heat, they turned off the computers. They knew that many physicists would gladly freeze to death if they thought they could get prime computer time.

By the end of December, the UA1 physicists had concocted a better way to prove whether an event was really a W or just looked like one. In Paris over the vacation the Saclay physicists, particularly Michel Spiro, the spokesman, and Denegri, had finally figured out the details of how to prove the existence of the neutrino in the W decay. They created a program that would add up the energy deposited throughout the nearly hermetically sealed detector and calculate if there was an imbalance. Simple Newtonian physics dictates that every action must have an equal and opposite reaction; hence, the energy deposited from the collision on one side of the detector had to balance with that on the other. If it didn't, it meant something had escaped detection, and the only thing that could escape would be a neutrino. If the missing energy, as it was called—the energy of the neutrino—when added to the energy of the electron in the collision, equaled the expected mass of the W, then the probability that the event was a W became overwhelmingly great.

The detector had been designed to completely surround the collision point for just this reason, to prove the existence of the neutrino by proving that energy was missing. Now they really could prove that they had W's. Real W's that even Richter would believe in. They had only five of them. But they sure as hell appeared to be real.

Rubbia had spent Christmas with his family, "looking at the Pyramids and sailing the Nile." It was his longest vacation in a decade, and the first time he had spent so much time with his family in probably twice that long. He called it decompressing. Before he left, Denegri had called him with the news that they could prove the existence of the W's.

Rubbia's only concern was whether they could make a convincing argument to the physics community. Whether he could overcome his track record. People had not forgotten the Alternating

Currents and the high-y anomaly. And if anyone was going to remind him, it was the CERN theory division in their satirical pre-Christmas play. This time it was an opera. In it, the hero, named Don Carlo, bursts onto the stage carrying his ppbar colliderscope, and a team of experimentors gathers around him. As the experimentors begin to claim that they have discovered a W, Don Carlo silences them.

"Shutta up," he screams at them. "I weela not ava you spreading de premature rumore!"

The experimentors look on, shamefaced, while Don Carlo goes on to announce that *he* has discovered the W. As he makes his announcement, he picks up a large flash card with the left half of a W on it, and then a second card with the right half of a W on it. He puts the two cards together to show the audience that he has discovered the W. He then turns one of the W cards over so that when placed against the other it makes a large Z.

"Wiza more sophisticated modern stateestical analyses," Don Carlo continues, "we alsoa find . . . We ava a Zed!"

Then he flips the cards yet again, and they make a Y, and he holds it over his head.

"What is thisa Y doing so 'igh?" Don Carlo asks. "Probably just an anomaly!"

Rubbia saw the play. "I saw Rubbia walk out of the room," said Kate Morgan, of UA1. "He didn't just walk out, like, time to go home, I'm finished. He slumped out. He sneaked out. And for weeks thereafter you could not mention the theory division Christmas play in front of Rubbia. It hurt."

On January 12, 1983, Rubbia and Darriulat were scheduled to speak on the results of the previous fall's run at a conference in Rome, which had been organized simply to air the new results from the collider. It was a showcase for CERN and Italy.

At UA1, the week leading up to this conference was madness. The analysis went on nonstop until the morning it began. There was the ever-present fear that UA2 might have beaten them to the W and that they had no way to tell. Or even worse, that UA2 would come out and say that there was no W. Even if the UA2 people were wrong in assuming that the W did not exist, they would ruin the impact that Rubbia was looking for, the bang. And what if they were right?

Even Sadoulet was working night and day by that time. He was doing a background analysis with Alan Norton, an ex-ISR physicist who had created the software for the central detector. They were trying to estimate what other possible, uninteresting factors might be responsible for what they thought were W's. What kind of junk might conspire to look like W's, possessing both the electron and, more importantly, the missing energy from the neutrino? The answer came out as close to nothing as they could wish. "Carlo left on a Tuesday for Rome," Sadoulet said. "I remember, we went to bed at four o'clock that morning. At seven o'clock, I was making the final computations of the background plots that Carlo took at one-thirty when he went to the airport."

Rubbia's talk in Rome was on jets, missing energy, and what he labeled "et cetera," and he told the physicists that he would concentrate on the et cetera. In an hour-long talk he never actually said that the W had been discovered, but he talked all around it, and he showed the events that were most likely W's, and that the background in them was infinitesimal. His message was clear, despite his caveat that the evidence was preliminary.

Darriulat followed. He, too, showed events that appeared to be W's—in fact, UA2's results were almost identical to Rubbia's—but his conclusion was less certain. "The need for more statistics is evident," he said. Rubbia had seen all he needed to see, however. UA2 would not deny the W's existence. The events that Darriulat had shown confirmed Rubbia in his own mind that he was right. "After we heard the UA2 talk," Cline recalled, "we were having coffee and Carlo said, 'We got it. We got it made.' "

Leon Lederman, from Fermilab, gave the plenary talk at Rome. His first point was: "The speed with which the data was analyzed and physics presented was truly astonishing, considering the complexity of the collisions, the sophistication of the detectors, and the hordes of experimental physicists." One of his last points was a statement on the value of the friendly competition between UA1 and UA2. It reminded him, as quite a few things do, of a story. This one was about two physicists, whom he called Carlino and Spierre, who were walking in the woods when they were confronted by a huge and fierce bear. "One of them (which one?)," Lederman related, "said to his colleague, 'A bear! Let's run!' The other responded, somewhat pedantically, 'You can't run faster than a bear.' To which the first physicist replied, 'I don't have to run faster than the bear. I have to run faster than you.' "

*   *   *

The CERN management had scheduled two seminars to present the new results. UA2 was scheduled for Thursday, January 21st, and UA1 was scheduled for the day before. It was at that Wednesday seminar that Rubbia officially announced the discovery of the W. In so doing, Rubbia had to present his evidence to the physicists of CERN; he also had to deal with his image, and the memory of too many mistakes, on the basis of which physicists were perfectly willing to disbelieve whatever he had to say. He had to summarize in one hour the culmination of almost a decade of working and hustling, leading and believing.

The main auditorium at CERN, which seats five hundred people, had maybe twice that many in it. Physicists shared chairs. Many could not take notes because they had nowhere to place a notebook and write. The doors were left open and people stood outside and looked over the heads of the people standing in the doorways. They sat on the grand piano in the corner of the auditorium, and even under the piano.

Rubbia was nervous. He sipped at his water, he pulled on his tie, he ran his hand through his hair and fiddled with his hundred or so transparencies. "He knew," said one UA1 member, "that he couldn't just go out and say, 'I'm Carlo Rubbia, and I say that this is a W event.' " He had been asked by the CERN management not to announce the W discovery in Rome, partly because they wanted it announced at CERN and partly because they didn't yet trust it.

Instead, Rubbia sold CERN the W with the passion that he had sold them the collider idea eight years earlier. He demonstrated how they understood the UA1 detector, how they know that it worked, how the central detector worked, how good was its precision, what its faults were. He anticipated every question, every criticism. When he was done with the apparatus, he told them about the physics, the evidence for the W, and why they believed in it. This is what we have done, he said, this is what we have seen, and this is why we must be right.

When he was finished, they gave him a standing ovation. Physicists who had made fun of Rubbia and predicted, even hoped, that his project would go down and take half the lab with it, physicists who had fought with Rubbia and swore never to work with him again, all clapped for five minutes. CERN was happy. It was beautiful physics, and they acknowledged it. It was their discovery, their vindication. They had waited nearly three decades for it. It was their Nobel Prize. They were euphoric. And the applause was for

Rubbia because for perhaps the first time in his career he had justified himself.

When, two years later, the Nobel Prize did in fact come, the physicists would all say, "Yeah, we know. Big deal. But it can't touch the W talk itself." After that day in the CERN auditorium, the Nobel was only a formality. Everyone knew then that Rubbia had pulled it off.

The following day, DiLella presented the evidence from UA2—in the same auditorium, to a slightly smaller crowd. The physics was virtually the same, but what would have been a tour de force in its own right was now simply confirmation physics.

On the Saturday morning, Rubbia joined Cline for a cup of coffee in the CERN cantina. There they met DiLella, Allan Clark, and Peter Jenni, all UA2 physicists. Rubbia looked unusually serious. He proceeded to tell the UA2 physicists that although he was convinced that both collaborations had what appeared to be W's, they should think twice before publishing. If it wasn't the W, Rubbia said, it would be the end of their careers. "And so," as DiLella explained later, "Carlo said that he had decided that he wouldn't publish."

In fact, Rubbia had decided nothing of the sort. The previous day he had delivered the first draft of his paper to Klaus Winter, a CERN physicist and editor of *Physics Letters,* the European physics journal. He had told Winter that the final draft would be quickly forthcoming, and that he would appreciate immediate publication. Then, saying that he had already alerted the journal, he convinced his group leaders to allow him to hasten the writing of the paper. But his physicists fought against rushing. Sadoulet argued that this was an important paper in the world of physics, and that they should not churn it out. It should be checked and thought over carefully.

"But it was important," Rubbia said later. "We were close. Had we waited three weeks, our priority claim would have gone to hell."

By Sunday the final draft was written. On Monday morning Rubbia gave the final draft to Winter, and later that night he sent the paper by courier service to Amsterdam to be hand-delivered at the offices of *Physics Letters.* Probably no more than a handful of the 135 UA1 physicists read the final draft. "People were shown a draft on Friday afternoon," said Eisenhandler, "and told that they had until Monday to comment on it. And when they turned up on Monday, they found that the paper had already gone out."

The UA2 physicists, on the other hand, wrote their paper in the conventional way and circulated it to all sixty physicists on the collaboration, who had two weeks to read it and make comments. It was published one month after the UA1 paper. Rubbia had established his priority, along with page one of *The New York Times*.

The episode in the cantina clinched the Nobel Prize for Rubbia, or at least according to Dave Cline. "We didn't know whether UA2 would try to send the paper in simultaneously," he said, "but they don't operate like that. It's not that they're super-honest, they just don't know how to move fast. They didn't think we would move so fast. It was a mistake, of course, because Carlo always moves fast."

Even two years later, physicists around the world assumed that somehow UA1 had discovered the W three weeks before UA2. As Ting put it: "I think there is no doubt whatsoever, at least in my mind, that the correct first published result came from Carlo. When you are beaten, when you are number two, you have all kinds of excuses. What is important is the published results." (Although most physicists never doubted that if Rubbia was to eventually get the Nobel Prize it would not be contingent on any worrying about unambivalent priority claims. After all, if tradition held, the prize would be for the collider, more so even than the physics or the detectors, and the collider, as DiLella said later, "was Carlo, and to a lesser extent van der Meer.)"

Still, Rubbia had not been alone in wanting to clinch an unambivalent priority claim. Herwig Schopper, the lab's director general, wanted to assure the Nobel Prize for CERN as well. He had been vacationing in Japan during the Rome workshop and the UA1 and UA2 seminars. He returned the following Sunday and immediately set about establishing priorities. In the interim, Erwin Gabathuler, Rubbia's old rival, had been in charge. Before Rubbia's W seminar, CERN had held a press conference, issuing a release that announced rather vaguely that experiments at CERN "begin to reveal the expected signature of a long-sought particle of matter: the 'W intermediate vector boson.' " No names were mentioned.

Schopper was much more direct, however. When he returned, he sent a Telex to his fellow lab directors around the world announcing that they had seen events at the collider: "Most straightforward interpretation is decay of W into electron plus neutrino." Then he went on to say that "UA1 has submitted paper and Rubbia is to present results at American Physics Society Meeting New York, this week."

Schopper called another press conference on Tuesday. The reporters who had been at CERN one week before were now told that the W had definitely been discovered. This time, the press release announced "A Major Step Forward in Physics: The Discovery of the W Vector Boson." It went on to say that since the last press release, three new developments had come to light: that UA1 had written a paper and sent it off; that the paper "confirms the discovery of the W intermediate vector boson"; and that "The discovery will be announced tomorrow . . . in New York, by the UA1 Collaboration Leader, Prof. Carlo Rubbia."

UA2 had been taken by surprise throughout the fall. As one UA2 physicist put it, "UA1 announced that they had a W early, and it was a crap event. But then things started to get out of hand. Each time they got an additional event that looked a little bit like a W, they said, 'Well, we got a W.' " All of the UA2 W events had come early in the run. DiLella had been analyzing the data, and even before Christmas he had what he considered a good signal. His colleagues did not believe it, however. Through January, DiLella argued that their claim was convincing. While Darriulat was telling the management that they needed more time, DiLella was putting together a seminar in which he was going to say that they had W's. Darriulat was conservative, a solid physicist. The physics was important to him, not the competition. DiLella also cared about the physics, but he wanted to win.

DiLella, who was one of the most likable characters in the field, was also not above trying to get revenge once they had conceded the W. The next piece of work on the agenda was to nail down the Z's, which the physicists had not expected to be able to do until the upcoming spring run when the collider would be tuned to produce five times as many collisions as in the fall. Surprisingly, however, UA2 had actually seen one Z event in the fall. The Z's were supposed to decay into, among other combinations, two high-energy electrons with a mass that added up to around 92 GeV. UA2 had spied an event with one good electron, and a second that had hit a coil in part of the detector and shattered into debris, so that all they could see was the debris. The mass was well within the range of a Z. Darriulat flashed a transparency of this event at Rome, but did not dare at that stage to call it a Z.

DiLella talked to Jack Steinberger about it. Steinberger had been Rubbia's adviser at Columbia, and had worked with him for several years at CERN. But then he had broken with Rubbia, and now the two never talked. Steinberger was also considered one of the best experimental physicists in the world, cautious and analytical. "He doesn't allow any kind of crap to get under the rug" is how one of his colleagues described him.

I told Jack about this famous event," said DiLella, "in which one electron was good and the other appeared to have showered. And Jack Steinberger said, publish it immediately. That is enough. If it is only one isolated event, if it has a mass of 94—even if one leg is sick—you should claim it is a Z-zero." DiLella relayed this opinion to his colleagues, and suggested they publish, but they refused. They wanted at least one perfect Z.

The next run was scheduled to begin in April, 1983. Early in February, DiLella decided that the only way to beat Rubbia in a race was to jump the gun. Two months before the run began, DiLella wrote a draft paper claiming discovery of the Z, leaving blanks in it for the characteristics of the events he hoped they would find. As DiLella figured it, they wouldn't need more than one day from the time the Z arrived to the time they filled in the blanks and sent the paper off to *Physics Letters* in Amsterdam. Not even Rubbia could move that fast.* "It was written," DiLella explained, "because we knew that we had been screwed so far. So let's screw him."

But it was Rubbia's year, and he was, so to speak, unscrewable. The spring run began on April 12. At five in the morning on May 4, UA2 got their first Z. They had programmed their computers so that they would literally ring a bell as soon as the detector spotted something remotely Z-like—in other words, the two electrons. That morning the bell rang. They looked at the event, and indeed, it was a Z. But, like the earlier one, it had problems. Although the mass was right around what it should be, this time both the electrons were disappearing into cracks in the calorimeters, and could not be measured accurately. "The paper had been written," said DiLella, "as though we really had gotten a golden Z-zero. So we said, let's wait for the next Z-zero, perhaps it will be golden."

By the time the UA2 physicists had decided not to publish, they had lost again. At three that same morning, Marie-Noëlle Minard,

*During the next run, one of the UA1 computer jocks, curious as to whether UA2 had found the Z yet, tapped into UA2's files and found the histograms from the anticipatory paper. Histograms are charts used by physicists to plot various quantities, and the ones he found were labeled "effective mass of the electron pairs." "It was obviously a Z histogram," said Astbury. "He was so bloody nervous about what UA2 was doing that he decided he had to have a look at this histogram. He found out how to do it, and was fantastically relieved when he found it was empty."

a thirty-six-year-old French physicist, was working late at her office in Annecy—a picturesque resort town an hour's drive from Geneva at the base of the French Alps. She was hoping to finish up a computer job before calling it a night. Minard had a printout of the collisions created in the UA1 detector a few days before, which had not yet been scanned. One event, according to the printout, had two high-energy electrons. She eyed the numbers for about a minute before deciding it was a Z-zero. Minard asked a friend, who was also working late that night, to drive her to CERN. They climbed into his old BMW and took off through the French countryside.

They arrived at CERN at five in the morning. Minard put the event on the Megatek, and there were those beautiful electrons. She called Jim Rohlf, an assistant professor from Harvard, who was working a late shift at the control room, and asked him to come over and look. "Sure enough," Rohlf recalled, "we really discovered it. We took a couple of hundred pictures of it, because we liked it. By that time it was about seven o'clock in the morning. I said, 'Well, Carlo's probably getting up now. You'd better call him and tell him that we've discovered the Z-zero.' But she didn't want to wake him. So I called Carlo up and told him we'd discovered the Z, and he said, 'Great, I'll be right in.' "

Rubbia lived a half hour from the lab. He was at the scanning room within forty minutes of the call. "We looked at the event for about two seconds," said Rohlf, "and he said, 'That's a Z-zero.' He knew. It was obvious compared to the shit we'd been looking at that that came the closest so far."

That event is now referred to as the famous Z-zero. Not just the first, but the famous one. And it was beginning a long and distinguished career in the physics papers of journals on both sides of the Atlantic. Like the UA2 Z's, it was not gold-plated. One electron was perfect, the other was not. The UA1 central detector had a magnetic field built into it, and the physicists identified electrons by comparing the energy of the track, as measured in the electromagnetic calorimeters, to the momentum, which was gauged by the curvature of the track in the central detector. If the two were the same, then they had an electron. The famous Z-zero had one electron in which the energy and momentum matched, and another in which the energy was 41 GeV, right in the Z-zero ballpark, and the momentum was 6 GeV. It failed the crucial test for an electron. It was Sadoulet's central detector that was responsible for the momentum measurement. When Rubbia saw the disparity, the first thing he did, as

Astbury remembers it, was to try to blame Sadoulet. "Hrmmph," Rubbia said, somewhat in jest, "it's Sadoulet buggering about again. We'll soon fix that." But in fact, whatever it was, it was not "Sadoulet buggering about again."

From the scanning room, Rubbia went to the cantina, where he flashed the pictures of the new Z to a few of his British physicists over coffee. "He's got this picture of two beautiful electrons going off," recalled one of the physicists. "And he's showing it to us like it was some kind of dirty picture. 'It smells like a Z,' he says. 'I know it in my bones. That's it. You can go.'"

Rubbia also called Herwig Schopper to break the news. But the director general was at a conference in San Remo, Italy, a conference loaded with the big names of physics. The message was relayed to San Remo. By nine-thirty that morning, on the strength of a phone call, Schopper had announced that the Z had been discovered.

That was it for UA2. By the time they got to CERN for their morning coffee, their yet-to-be published Z-zero paper was redundant.

Throughout the following week, UA1 physicists ran test after test on the detector to find out what might have caused the mismatch in that one electron, and they found nothing. The next week, Rubbia showed the one Z event to physicists at Princeton, and a few days after that it was in *The New York Times*. The paper quoted an American physicist as saying that the odds were three to one that the event was a Z, and that "someone like Dr. Rubbia could not be expected to stake his reputation on one event."

Rubbia had indeed argued to publish on that one event, but the collaboration fought against it. "We were in the following dilemma," Sadoulet explained. "Carlo was going around saying, don't worry about this track in the central detector. It is not well measured. Sadoulet does not know what to do with the central detector. But Gabathuler [the research director] and other people said, if we can't believe your central detector, then we don't believe your W story. But if we do believe your central detector, then it may not be a Z."

On May 27, Rubbia officially announced the discovery of the Z in a seminar at CERN. By then UA1 had one more Z, which was indisputable. On June 1, Rubbia mailed off the paper to *Physics Letters* and CERN distributed its official press release. Since the preliminary announcement had already been made in *The New York Times* and then carried throughout the world, the newspapers treated the new press release as somehow confirmation of an earlier discovery: "Physicists Confirm Discovery of Z-Zero," said the *Times* headline. By now UA2 was irrelevant.

Still, the competition was not over yet. At the time of the seminar, UA2 had not yet found their third Z. By the end of the run, July 4, they had eight. But they said nothing. Four of the eight were gold-plated Z's, and four were funny. Many of the UA2 physicists, Darriulat among them, were worried that they were seeing some kind of bizarre decay of the Z, an unexpected, unpredicted, and hence very important decay. They also knew that Rubbia had already taken the credit for the discovery of the Z, and that they had nothing to gain by hurrying; they would wait.

The analysis was now being done by Tom Himel, a young American physicist who had learned his physics at SLAC. "We had made the decision that we might as well wait and do it right," said Himel. "We would look like real fools, if we go and confirm Carlo on the basis of evidence we're not quite sure about, and then later on decide that the electron is always accompanied by something and that's what makes it look bad in the detector." Of course, he admitted, they also hoped that something funny was happening, and that by proving it, they could still make Rubbia and UA1 look foolish.

Meanwhile, the CERN management was getting nervous. As with the W, the management had wanted as much as Rubbia to establish a clear priority claim to the discovery of the Z and to make sure that it was known that this discovery was unambiguous. They didn't want another fiasco like that over the Alternating Currents. Day after day, the UA2 physicists were pressed in various meetings to tell whether they had Z's, and day after day they said nothing. By now, the management feared that they would contradict Rubbia's results. Then, to ease the pressure on his group, Darriulat officially announced that UA2 would say nothing until the run was over on July 4.

By then Himel had finished his background analysis and found that Rubbia and UA1 had indeed been right. On July 7, Darriulat held his seminar, demonstrating that UA2 also had the Z. (But not before the CERN management had privately seen the UA2 results so as to obviate any potential embarrassment.) When the management announced the confirmation in a press release, they did not ask UA2 to propose the wording—as was the tradition—and refused UA2 even the opportunity to question the wording. They had had it with UA2. The CERN press release was unequivocal: Z-zero discovery confirmed. Results in excellent agreement with the published UA1 results . . .

Before Herwig Schopper had taken his position at CERN, he had been in charge of the Deutsches Elektronen-Synchrotron Laboratory, DESY, at Hamburg. It had been his lab in 1979 when gluons had been discovered there. Gluons were in some ways as important a discovery as the W, but the announcement had been mismanaged; somehow the physics community never really bought it. There was no bang. Schopper wasn't going to let that happen again.

The DESY accelerator had four major experiments, and all four saw the signs of gluons and published quickly. The discovery was first announced at a conference at Fermilab, however. This prompted European physicists, who thought that the evidence was not at all conclusive, to question why it had been released first in the States and to answer, as the *New Scientist* wrote, "that American particle physicists are trying hard to keep up the momentum for federal funding for their expensive form of research."

In addition, Ting had one of the four DESY experiments, and he had used extreme tactics to establish priority. Ting had sent his paper off to *Physics Letters* with a title claiming the discovery. The journal promptly rejected it. As Klaus Winter recalled: "I said, you couldn't say that. It is not your right to call this a discovery. It was up to the community to judge whether this was a discovery. It was an observation." What Winter did not tell Ting was that he had already received gluon papers from at least one of the other experiments. Ting then hand-delivered the paper to the American physics journal *Physical Review Letters*—located at Brookhaven, where Ting had discovered the J/psi—and they published it immediately.

By the time the whole affair had ended, as Don Perkins told me, "It was pretty unpleasant. Just a lot of people trying to get a lot of publicity."

For the W and the Z, Schopper was not about to make the same mistake. He had begun, with Rubbia's help, to sell van der Meer and Rubbia to the Nobel Prize Committee. The Nobel Prize could be given to at most three people. One of the faults with the gluon discovery was that too many principals had been involved: four experiments, four group leaders, and four sets of results, all nearly identical. Schopper had tried to back Ting's priority claim in various physics publications, but it was a hopeless task. The same went for the neutral-current experiment at CERN a decade earlier. Nobody could decide who should get credit, and the press releases announcing neutral currents mentioned no names at all. It was a faceless discovery.

This time around, every press release came with a supplement that explained how this was all made possible by stochastic cooling, courtesy of van der Meer; and no press release, even the one announcing UA2's confirmation of the Z, was without Rubbia's name.

Behind Schopper's urgency to get the Nobel for CERN was the Large Electron Positron machine, which was Schopper's baby. He had already procured the money from the participating European governments to build it. Although he had had to accept a scaled-down version of the machine—only half as powerful as originally designed. He would still have to fight every year to keep the money coming in, and do it against a grim economic background in Europe of declining outputs, inflation, and unemployment. LEP was a $500 million project according to CERN's bookkeeping, which didn't include such things as shop costs and labor provided by the lab, and nearly a billion dollars according to U.S. bookkeeping, which did include these things. Schopper dearly needed some international prestige, and the Nobel would provide it.

At Rubbia's press conference announcing the Z discovery, a month before the collider run ended, Schopper told reporters that the discoveries were the most important in physics since the invention of the transistor twenty-five years earlier; they put CERN at least six years ahead of all its competitors, and they would surely merit a Nobel Prize in physics. It was one of the few occasions in the history of science that a lab director had told the press to inform the Nobel Foundation where his vote lay—before the experiment was even finished.

When it was over, the physicists professed that the practice of their craft had seemed to change along the way. The competition had somehow become more important than the physics. It was partly the money: The collider and the two experiments had cost in the neighborhood of 200 million Swiss francs (roughly $100 million), and with that kind of investment, the laboratory and the physicists had to produce results. And it was partly Rubbia, who would be damned if he would let anyone beat him to anything.

The tradition of agonizing for months over a physics paper or a result of importance was ended. When Ting had discovered the J/psi he had spent six weeks checking his equipment before he would publish. And that J signal had stood out from the background like the Eiffel Tower. Then he had announced it because he knew that Richter had it too. When, in 1975, Nick Samios discovered the first charmed baryon* and published on one event, he did so after six months of checking the backgrounds.

The first W event that Rubbia took around the world had been in his hands for maybe twenty-four hours. And it would turn out not to be a W. Two years later, however, Rubbia still argued that it was. "It's a fine W, except it's not a perfect W," he told me, even though he must have known it was considered background by all his physicists.

Sadoulet provided one possible explanation. "You have to see how Carlo is working," he said. "Carlo is first a political being, and will say things even if he has no confidence that they are true. Just for the political purpose." The political purpose is to establish priority. Michel Spiro, head of the UA1 Saclay contingent, suggested that it was conceivable that none of the UA1 physicists ever bothered to tell Rubbia that the event had been proven not to be a W. And that even if Rubbia had noticed, he would not have wanted it discussed in public, because he would not want the world to know.

The non-W had served Rubbia's purpose, all the same. Even two years later, CERN physicists would point out to me, incorrectly, how careful Rubbia had been with his W. They would say that in October and November of that year, Rubbia could be heard in the cantina at all hours of the day talking about his discovery, but that still he waited until January before publishing.

Rubbia seemed to understand that whether it was a W or not was almost inconsequential. What mattered was that it proved that UA1 could discern W-like events. Since the theory, which was almost certainly right, predicted that more W's would be on the way, UA1 had every reason to be optimistic. Whether it was a W or not, Rubbia could use it to claim the discovery; as long as those other W's showed up, no one would ever know the difference, or even care. Added to this, Rubbia seemed simply unwilling to waste his

---

*Baryons are particles, like protons, with three quarks.

time with background. Background calculations did not win Nobel prizes. He let his junior colleagues worry about background.

Rubbia once directed me to the physicist who had personally handled the W analysis, saying that he could explain why that well-traveled first event was a W. When I asked him, however, this physicist explained why it was not. But he did say, "If the event satisfies nine out of ten of your requirements, then you have to get your event out into the public view. And you have to be honest about it. You can't sit on something that exciting, that pressing. The whole world was waiting for those events."

And that was one of the problems. The whole world was waiting. The UA1 and UA2 physicists talked about living in a fishbowl, the eyes of the world staring in at them. Darriulat had complained to me that the world had lost its critical judgment. Nobody wanted to understand the details of the experiment, they only wanted the results. "You can say almost anything," he said, "and you don't feel among the physics community a strong desire to understand. What really interested many people was whether you had a gold-plated event or a 'crappy' event. That was the kind of slang that was used without giving meaning to these words." He added that it was at its worst among the CERN management.

In fact, outside UA1 and UA2 most physicists simply could no longer understand the two experiments. They were too big and too complicated. In the old days, everyone knew how a bubble chamber worked and how to analyze the pictures that came out of it. Anyone in the field could judge the worth of a journal article. Now, even within what Sadoulet calls the "microfield of colliding-beam physics at 540 GeV," it was difficult to know what was good and what was bad. The UA1 central detector was a marvel of technology and physicists ogled the pictures that came out of it. But they didn't necessarily understand what it was telling them. They took Rubbia's word for it. And Rubbia was using his nose, saying, "It smelled like a W," or, "I feel in my bones that it's a Z." It even took the UA1 physicists an extra six months before they figured out what the undeniable flaw was with that first W—the one that was hanging on the wall in the Department of Energy in Washington—and why it couldn't be a W. (When Rubbia showed the first W, which was not a W, to Weinberg, Salam, and Glashow in November, 1982, they all apparently agreed that it seemed to be a W.) But still it went in the paper. As Anne Kernan, the leader of the University of California at Riverside group, told me, "It was knowingly included to strengthen the statistics." And nobody could tell or seemed to care, with the possible exception of Richter.

This was also one of the few times in physics that the only two experiments in the world that had a chance to win were both on the same accelerator. Before it had always been Fermilab and CERN, or SLAC and DESY, or Brookhaven and SLAC, or some combination of all of them. Now, it was CERN and CERN; the same machine, the same energy, the physics equivalent of man on man. At times the competition was felt so intensely that it overwhelmed the science.

The physicists viewed all this in one of two ways. At UA1, for example, Anne Kernan said that she admired Rubbia's courage for speaking out when he had the one Z event, and planning the press conference when he had only two events, even though both had problems. On the other hand, at UA2, Tom Himel could not believe that they had written the Z paper complete with blanks in anticipation of the event. "It's a big responsibility, publishing a paper," he said, "and saying you've found something and you've done it and it's right. The first Z came in, and we should have been really excited. But really my first reaction was 'Oh, no, now we're going to have to decide whether we're going to publish on the basis of one event.' "

Rubbia's collider had changed the balance of power in the world of physics, as well. He had proved not only that physics, very good physics, could be done with protons on antiprotons, but that the W and the Z existed and that the standard model was correct. Perhaps more important still was that he had proved that Europeans could do physics as well as or better than the Americans.

Six days after the Z-zero press conference, *The New York Times* ran that classic physics-as-competition editorial: "Europe 3, U.S. Not Even Z-Zero." The editorial suggested revenge.

Only a week later, an advisory panel of American physicists voted to cancel construction of Isabelle at Brookhaven—the machine that, to some extent, had kept the collider from being built at Fermilab, but that had been limping along. Instead of Isabelle, the panel recommended that the United States go full speed ahead on the Superconducting Super Collider (SSC), a massive machine built along similar lines to the CERN collider, but a hundred times more powerful.

This machine was as much a reaction to Rubbia's success as anything else. It was an unprecedented decision. "They terminated a project which had overcome its technological problems," said former SLAC director Wolfgang Panofsky, "and instead recommended a project that did not yet have a proposal, a director, or a site." No one would deny that the SSC was not a machine that would do great physics, but the idea reeked of revenge. For the first time since World War II, the American physicists were number two in high-energy physics, and they did not like it.

Further, it appeared that the situation was only going to get worse.

8.

BEYOND THE END
OF THE ALPHABET

"Carlo really wants to know more than anybody else, and he wants to know
before everybody else . . . I don't see any end to his quest."
—Antoine Leveque, co-spokesman of UA1

Physicists have been known to say modestly that they cannot believe that such a lowly creature as man could have learned so much about the workings of the universe. Then in the same breath they will talk about creating still more elegant theories, and about all the problems with their present model. Their three working theories—relativity for gravity, quantum chromodynamics for the strong force, and the Weinberg–Salam model for electromagnetism and the weak force—they refer to as the "standard model." They work, and they work well.

Nevertheless, within the standard model there are seventeen genera of particles. Physicists consider this an inelegant and aesthetically unsatisfactory number. They would prefer one, or two, or maybe three. And the model has roughly thirty arbitrary parameters —such as the respective masses of these particles—that have to be measured in the labs and then penciled into the equations by hand; nowhere in these theories does the mathematics decree that these arbitrary parameters must be what they are. Then there are the two separate quantum theories and an entirely different theory for gravity, and that doesn't quite sit right either.

For a century, physicists have believed in their hearts that the workings of the universe should be extraordinarily simple; that all the forces should work in the same way, no matter how little they appear to do so on the surface; that they should all be descendants of one primeval force that ruled at that instant of creation. Their ultimate goal is to write one theory that would include every force and every particle. And they are working toward that goal one step at a time.

The best candidates around for this next great theory of nature are called grand unified theories. These theories attempt to link the quantum theory of the strong force with the quantum theory of the electroweak force. They predict that there will be no new particles discovered that are heavier than the W and the Z (except maybe one enigmatic particle known as the Higgs) until physicists learn how to build an accelerator as big as the galaxy and hit energies a trillion trillion times higher than those in Rubbia's collider. They call this possibly forever uncrossable expanse of energy "the desert." If there is in fact no life in the desert, no new particles, then there will be no more evidence forthcoming with which to build better theories. Progress will be at an end. The standard model will remain standard for the duration.

Dave Cline, Carlo Rubbia, and Peter McIntyre had something to say on the subject of the desert when they wrote their very first proposal for their collider in 1976. "The collision of intense beams of matter and antimatter at extremely high center of mass energies," they wrote, "is more than likely to involve completely undreamed of physics." This scorned the predictive powers of the theorists. But Cline, Rubbia, and McIntyre were selling the proton-antiproton collider, and they were willing to say just about anything to get people hooked. Then again, they were experimentalists, not theorists. They went on to say that it would be of no use whatsoever to try to write down what they might find at these high energies. "The most interesting discovery would almost certainly not appear on any list made at this time," they added. "This is, in fact, the main reason for building the machine."

As expected, no one really listened to them. They were simply repeating what had become one of the unwritten canons of physics: Build a bigger machine, get to energies that no other machine has reached, and you'll find something new. Physicists loved to cite the classic example of this, the Berkeley Bevatron. This machine was built in the fifties to find one particle, the antiproton, whose existence had been predicted twenty years earlier by Paul Dirac's quantum theory of electromagnetism. The Bevatron was switched on in 1954, and it found the antiproton in 1955, resulting in later Nobel prizes for the physicists concerned. Then the Bevatron started to turn up one new particle after another. These particles were called resonances, and they are very unstable, existing for the briefest moment before decaying into more stable particles. By 1959 the number of particles in the physicists' pantheon had doubled, and would more than double again in the next four years.

This argument was always the ace up the physicists' sleeve when they built any new machine. "Frankly," Rubbia once said, "when we did this thing, we didn't know what the hell we were doing. The only thing we did know at the time was, we got more energy than anybody. We can't lose."

Not more than two weeks after UA2 announced that they had confirmed the Z discovery—the last of the planned discoveries using the collider—Rubbia started talking up the unplanned discoveries. In Darriulat's seminar on the Z, the Frenchman reported that one of their worrisome Z's was undeniably a bit strange. It had the two electrons that were the signal of the Z decay, but in addition it had a photon (the particle that transmits light and electromagnetism). The mass of all three together was equal to the mass of the Z. It was as if somebody had thrown in an extra particle that nobody had ordered. Such a configuration could exist, according to theory, but it would be exceedingly rare. UA2 wrote it up as a freak occurrence. Darriulat showed it at the seminar, and that was that.

When Rubbia saw this, he rewrote the script. He and his colleagues claimed that their famous first Z, which Rubbia had used to announce the Z discovery, must have been the same kind of freak occurrence. Rubbia also reported that they had another Z with the same accompanying excess photon. That meant that in three Z events out of a total of fourteen, they had these strange radiations. The standard model, so Rubbia reported, decreed that the odds of that happening were one in ten thousand. To Rubbia, this suggested that maybe the theories were wrong. It was certainly premature to draw any conclusions, he said, but "it shows how entirely new phenomena may be occurring in the collider energy range."

The UA2 physicists were incredulous: They couldn't believe that Rubbia had taken the single event that he had heralded to claim discovery of the Z, and then suddenly reclassified it as potentially a new phenomenon as soon as it had become in his best interests to do so.

The theorists, on the other hand, had no problems with Rubbia's new claim. They went wild. This was what they had been waiting for. The interesting work had all been done for the standard model, and the Nobel prizes had been given. They needed something concrete that went one step beyond, or even a vague notion, on which to get working. Over the next six months, the theorists filled reams of papers with their speculations, covering everything from minor adjustments to the standard model to predictions that electrons were not elementary particles, as everyone had believed for the past sixty years, but were indeed made up of smaller particles the way protons and neutrons are made up of quarks.

Still, the data were simple, and neither theorists nor experimentalists could do much more than play with their models and wait until the next run to see if more of these bizarre events arrived.

So they waited. And within six months, UA1 let slip that they had misinterpreted their background calculations and that the odds of the existence of so many bizarre Z's were not nearly as great as they had originally announced. In fact, it was no big deal, they said, and the theorists could probably ignore it. By then the theorists had published over fifty papers. But they didn't complain. They had even more bizarre results to confront.

At the end of the spring run in 1983, Rubbia announced that they soon hoped to dig another discovery out of the data collected. That was the sixth and last of the quark family. The fifth had been found in Leon Lederman's upsilon discovery (the upsilon was a bound state of two quarks, known as bottom quarks) and the theorists figured there should be a total of six. Six was a nice round number; there were six leptons, so why not six quarks as well? The sixth was known as top in America or truth in Europe.

In fact, two American theorists, suffering from a virulent case of Nobel fever, had already claimed to have discovered the top quark in UA1's own data. They had seen in a W paper that UA1 had discarded background events that had the expected signature of the top quark. They wrote a paper to this effect. It was a less than a brilliant deduction, however. They had not noticed that these were events that had been thrown out by the UA1 physicists because the electrons in them were not real electrons, but other particles faking electrons in the detector. When Rubbia heard about the paper, predictably he blew up. The next time he faced reporters at a press conference, he confidently told them that he expected to get the top by year's end.*

*In June, 1984, Rubbia announced that UA1 had observed events consistent with the top quark. His claim was not definitive because the evidence was still inconclusive. On June 25, in an article entitled "Physicists May Have Tracked Last Quark to Lair," Rubbia told *The New York Times* that the evidence "looks really good." On July 4, CERN confirmed the discovery in a press release. "Laboratory Confirms Existence of Top Quark," the new *Times* headline read. The evidence had not changed in the intervening period. In October, UA1 admitted that the evidence was inconclusive. A year later, the evidence would be even more inconclusive.

*   *   *

At UA2, they knew that Rubbia would go after the top quark next, so they figured they would try to beat him to it. But the situation with the top quark was even stickier than that of the W and the Z. UA1 had the better chance of making sense out of it, but . . . Tom Himel, the young American, began looking for it in July, 1983, as soon as the Z was wrapped up.

Unlike the W and the Z, which had been surprisingly easy to find, the top would be mired in a swamp of background. If any could be found, they would come from W decays: The W, instead of shattering into an electron or a muon and a neutrino, would blow into a top quark and an antibottom quark. The top quark would then decay into an electron and a neutrino and a jet; the other quark would fragment into a single jet. The analysis was complicated, and maybe hopeless. Himel was justifiably not too optimistic.

After two months, he reported that if the top existed, he certainly couldn't find it. But he did find something else: two events that seemed to be inexplicable. They appeared to be W's—with the electrons and the missing energy—but they had a jet emerging from the side. And the total energy was huge, far beyond the range of the W. They simply did not expect W's with energies like that. Himel was stumped. By that autumn he had found two more. He kept working on it while the rest of the UA2 collaboration ignored it. As DiLella said, "We got these events, and we said what the hell are they? And we forgot about them." There was little else they could do, except wait for the next run and see if they got more. But in the meantime, the situation blew up around them.

In February 1984, Pierre Darriulat attended a workshop in Chicago. He had breakfast with some theorists who were doing what theorists love to do whenever they get the chance—speculating about what the next great theory of nature might be, and what it would look like if ever it showed its face in one of their accelerators. Darriulat casually mentioned that they had these four strange events at UA2 and did not know what to do with them. After that, he could only step back and get out of the way. As in Virgil's *Aeneid,* in which rumor "flies between the earth and heaven, shrieking through the darkness," the conference started buzzing. Word flew around the world in the time it takes to dial a telephone.

One of the people at that breakfast had a pet theory that W particles could be created in what are known as bound states—in other words, two W's locked together. He suspected that one of the

UA2 events fitted his theory well. No more than thirty-six hours later, Luigi DiLella got a call from an old friend in the States who said, "Hey, I understand you guys have got particles that decay into two W's." DiLella knew immediately what he was referring to, but he had thought that those funny events were still sitting quietly in one of the drawers at UA2. When Darriulat returned to CERN the next day, they gave him hell for talking about the events in public, even though they knew it wasn't really his fault. He had just misjudged the size of the vacuum in the physics world that would suck in news of anything even remotely strange. UA2 then decided to publish, thereby putting at an end the rumors and speculations.

A second theorist at the breakfast in Chicago had been working for the past few years on a candidate for an ultimate theory of everything. After he heard what Darriulat had said, he thought about it for a day, worked out the calculations on a blackboard, and decided that the funny events were easily explained by particles in his favorite theory. He naturally decided to ascertain whether UA1 had any of these strange events. Other theorists came to the same conclusion. Both Dave Cline and Anne Kernan of UA1 were at the Chicago conference, as was Rubbia, who arrived late. They were all approached by these theorists asking them if they had heard of the UA2 events, and if they had anything like them. What events? they said. And the theorists told them. The UA1 people replied that they didn't. Perhaps UA2 was wrong. Or they said they were not sure, and that they'd better check. But they said nothing about whatever their own physicists might be working on.

At UA1, Jean-Pierre Revol, a young French physicist, with a couple of colleagues had been wading through over two thousand computer tapes of recorded collisions looking for signs of physics beyond the W and the Z. He was particularly interested in events with huge missing energy. Several of the candidates for the next great theory, especially one known as supersymmetry, predicted the existence of particles that, like neutrinos, would pass through the UA1 detector without leaving a trace, and hence would appear as missing energy. (This work was also being pursued independently by another French physicist, Aurore Savoy-Navarro, but later her efforts would be overlooked in favor of Revol's work, which was done under Rubbia's eyes.)

Revol was thirty-seven years old, looked five to ten years younger, and had enormous energy, exuberance, and verve. He was

an expert rider, a fanatic runner, and a devotee of fine wines, which
he would order carefully for any meal from roast quail to pizza. He
also liked the idea of making his mark in physics. He had done his
Ph.D. under Ting at DESY and had managed to impress Ting with
his talents. Then he went to UA1. When I once asked him why, he
told me: "A lot of people told me not to go to work with Carlo
Rubbia. But why should I do crappy physics just for the sake of
being with nice people?"

At a meeting of the UA1 collaboration in London in January,
1984, Revol reported that he had some interesting phenomena—
three or four events that seemed to be peculiar. They could have
been W or Z decays, but they seemed to have too much energy. It
was not definitive, but it was enough to intrigue his colleagues, and
Rubbia.

At the time, Rubbia was trying to convince CERN to build an
extension to the antiproton accumulator to increase tenfold the
number of antiprotons it produced, and CERN was stalling, saying
it sounded too expensive. Rubbia was also trying to push through
an upgrade for the experiment that would make it more powerful.
He dearly wanted to be able to hand the CERN management at least
the first signs of something new, so that he could say, "See? Now
we have to be even better next year so we can make this crystal
clear."

Rubbia, who had been pessimistic about this work, suddenly
began to encourage Revol. He suggested that other physicists work
with him. His assistant professor, Jim Rohlf, also joined up, along
with Mohammad Mohammadi, an Iranian graduate student at Wisconsin, and a few others.

In mid-January, Mohammadi found the ultimate bizarre event!
It would come to be called Event A, as though it was evidence in
some higher court that decided on cases having to do with great
theories of the universe. He had been working late one weekend at
the Megatek when he found something that at first he thought was
a W. It had a high-energy muon and missing energy. Then he realized that this creature also had a huge jet around the muon, and that
the missing energy itself was huge, maybe twice what it would be
for a W. It made no sense at all. "I started looking at this thing,"
Mohammadi recalled, "and I noticed that it was very unusual. If it
was a W, then what was this jet doing there? And if not, what could
it be? I took some pictures of it, but I was so excited that I just
couldn't sit there. I got up to walk around and see if I could find

somebody to come and see this thing. In particular, I was looking for Jim Rohlf, because he was the kind of guy who wanted to see these unexpected things. I wanted to ask him to verify the thing. Was this crap that I was looking at? Why was it that nobody had seen this thing before?"

He found Rohlf in his office, working his usual hours, which went from noon until three or four in the morning. Rohlf was skeptical about this event, but he agreed to join Mohammadi in the scanning room to take a look. On the way there they passed Rubbia at the copying machine, and Rohlf mentioned that they were going to look at this strange event. Rubbia waved them away as though he wasn't interested.

Rohlf studied the event silently for more than an hour. He was trying to figure out how the detector could have screwed up and faked what they were seeing. But by two in the morning, Rohlf had no answers either. The event seemed to be inexplicable. The detector appeared to have worked perfectly. But if it had, then what the hell were they looking at?

A week or so later, after Event A had been shown to the collaboration, Rubbia gave his opinion of it. As Mohammadi put it, "Carlo walked up to Jim Rohlf and said, 'Hey, Jim, have you seen this event? Isn't it spectacular . . .' And Jim said, 'Jesus Christ, Carlo, we told you about it that night.' "

After Darriulat's indiscretion in Chicago, and after the rumors began their inexorable spread, life at UA1 went to critical mass. A collider conference was scheduled for the first week in March in Bern, Switzerland. Rubbia had no idea what UA2 was planning, although there were wild rumors of strange particles discovered by UA2 with masses twice that of the W. When Rubbia returned from Chicago, he ordered his lieutenants to search through the data for anything like UA2's funny events. They found nothing.

The week before the conference in Bern, they worked close to twenty-four hours a day, with last-minute meetings and arguments about what should be said and what kept quiet. It was like the week before the Rome conference the previous year, only worse; nobody really knew what they were looking at. They had some events that seemed absolutely remarkable, but at the last minute they realized that they had a bug in their analysis program and that those bizarre events were in fact junk, just boring background. But they still had

Event A, and it was still inexplicable, and they had four or five other examples that were almost equally weird.

At Bern, Rubbia showed those inexplicable events. Two had a single photon on one side and huge missing energy on the other. Then there were those, like Event A, that had jets on one side and huge missing energy on the other. He called them monojets. He said that they did not see the funny events that UA2 had seen, implying that because UA1 was the better experiment, UA2 must be wrong. He went on to say about his own stuff that "new phenomena have been observed which cannot be explained in the standard model."

After Rubbia, André Roussarie, of UA2, presented the strange events that Tom Himel had unearthed the summer before. A second UA2 physicist, John Hansen, showed the data that were the source of the other strange rumors, a particle at nearly twice the mass of the W. But they admitted that the signal was weak.

This time UA2 was going to take no chances. After the conference, they immediately started on their paper. They had heard Rubbia say that he had none of their strange events, but they knew how little that meant. It might have been a replay of that Saturday morning in the cantina when Rubbia bluffed them out of rushing their publication of their W events. "We had to get our stuff out and into the public eye fast if we wanted to somehow get credit for it," said one member of UA2. They also combed through their own data for monojet events like those found by UA1, but because their detector was not as comprehensive as UA1's, they had little hope of finding anything, and in fact they didn't.

The UA1 physicists spent a hectic two weeks trying to figure out whether they really did have a totally new phenomenon, now that Rubbia had already claimed they might. Then Rubbia was handed a copy of the new UA2 paper on their strange events. UA2 had mailed it off that day to Amsterdam to be published. Rubbia decided that UA1 would have to be in the same issue. "Carlo got the paper in his hands on either Wednesday night or Thursday night," recalled Witold Kozanecki, who had done his doctoral work under Rubbia. "I remember I went to dinner around seven o'clock and came back around ten. When I got back, Carlo was running up and down the corridor waving the [UA2] paper in his hand, saying, 'I want a paper by next Monday.' In fact, he wanted it by Friday night."

Thursday night Rubbia started writing. Not even his lieutenants really knew what he was going to say. Mohammadi remembers working late with Revol and Rohlf in the UA1 conference room across from Rubbia's office when Rubbia came down and said he was going to start writing the paper. "Fifteen minutes later," said Mohammadi, "he came back and said, 'How does this sound, Jim?' Most of the time he was talking to Jim. And he was just pacing the room, going back and forth, reading like mad. He read the abstract [the introduction to the paper that summarizes what will be said]. At the end of it, he just said this was new physics. As soon as he walked out the door, I asked Jim, 'Jesus Christ, what is he saying now?' And we just looked at each other; we just sat there looking at each other."

Rubbia had his first draft by Friday night. He sent out a notice to his executive committee that they would have a meeting the following Tuesday morning to discuss objections to the paper. Rubbia and a handful of his physicists worked through the weekend refining the numbers that would go into the paper. By Monday the paper was finished, and Rubbia got it to Klaus Winter, the *Physics Letters* editor at CERN, who made his criticisms and handed it back. They addressed the criticisms Monday night, and sent the paper off to Amsterdam Tuesday morning.

If any members of the collaboration thought the paper had flaws, they said little about it. Sadoulet, who had been in Brazil while the post-Bern rush was on, had tried to argue against Rubbia's insistence that they get into the same issue of *Physics Letters* as UA2, and, as he put it, "was sent back into my office." As for the rest of the collaboration, one UA1 member told me, "They had very little chance to even look at the damn thing."

When Michel Spiro, the head of the Saclay contingent, received the preliminary draft in Paris, his first response was "This is an incredible paper, I do not believe in it at all." The electromagnetic calorimeters had been built at Saclay, and they had spent the last two years learning to calibrate them. This meant gauging through a variety of mechanisms exactly what the difference was between the actual energy in a particle and the energy reported by the calorimeters. (It would take another year, still, before they professed truly to understand the responses of their calorimeters.) Spiro believed that most, if not all, of Rubbia's new physics could be explained by experimental weaknesses in his calorimeters. He sent a letter to the collaboration explaining his criticisms, but by that time the paper was out. Few of the physicists involved with the analysis ever read his letter. He then telexed a message to the editors of *Physics Letters*, requesting that his name be taken off the paper.

* * *

The UA1 paper became known as the monojet paper, and it made some interesting claims. It said that they could not "find an explanation for such events in terms of backgrounds or within the expectations of the standard model." The first draft had been even more dogmatic; it had said: "No conventional mechanism appears to be capable of producing such events, which must therefore be due to some new physical process." But Winter had suggested in his role as editor that this was too strong. In the final draft, they tempered that critical "must" to a "may." The final draft also suggested that the effects might be caused by new particles, and offered a number of theoretical speculations "that may be relevant to these results." It mentioned just about every candidate for the next great theory as a potential explanation for what they had seen.

Physics Letters published both the UA1 and the UA2 papers in the same issue. Rubbia had achieved his goal. Nevertheless, physicists complained later that they couldn't understand the paper; that it was sloppy, below UA1's standards. If it really was new physics, they had no way to tell. "I don't think the analysis of that stuff was done very well," said Burt Richter later. "It might be right; it might be wrong . . . it seemed to me it was quite premature to say that it was some new phenomenon." Several UA1 physicists were harangued at conferences. Sadoulet remembers being attacked by one prestigious theorist at a talk at Fermilab. "I won't speak about this paper," the theorist said. "It's crap. And it's well known that it's been published just for the Nobel."

That physicist knew Rubbia and the UA2 physicists well. But just as important, he had also noticed the line that betrayed Rubbia's ploy. Included in each paper in *Physics Letters* was the date on which it had been accepted by the journal. This was a convention that decided the more vicious priority battles. The acceptance date of the UA2 paper was three days before that of UA1's. The physics community had not missed this fact. They assumed, rightly, that Rubbia had rushed the paper because he feared that if they had discovered some spectacular new phenomenon, it would be bigger news than the W or the Z. And if UA2 discovered it first, then Rubbia might have to split three ways the Nobel Prize that he expected to get in six months. Or even worse than that, instead of giving him the prize that year, the Nobel Committee might wait for the dust to settle before awarding it, and that could take God knows how long.

* * *

A month and a half later, Rubbia presented his new physics again at a meeting of the American Physical Society in Washington, D.C. This conference is known colloquially as the Washington meeting. It has always been a favored occasion for announcing new results, since it provides physicists with the opportunity to impress the bureaucrats at the Department of Energy and the National Science Foundation—the two agencies that fund high-energy physics in America.

After his presentation, Rubbia told reporters that the findings of UA1 and UA2 on the W and the Z confirmed beautifully the predictions of the standard model. "Now we know why things happen the way they do," he said. "So you may worry about something else." Then he told them about that something else. Rubbia explained that his latest results might have something to do with that "gloomy view" of the desert—as postulated in the grand unified theories—of an enormous expanse of energy in which no new particles would be found. Rubbia described the monojets, and when asked how strong the evidence was for these bizarre events, he said, "Let me say there are more of these events than there were W's when we said we had the W."

He concluded that the "desert, so to speak, was blooming." Then, pushing that metaphor even further, he told of "the famous story of the man who was digging in the desert and saw a little pyramid underneath which was a caption saying that it might be the discovery of the century, it all depends how deep he goes."

They had seen these events, Rubbia explained, at the farthest limit of the energies that his collider could reach. It was as though they had looked through the most powerful telescope in existence, and had seen blurred and just barely visible, at the farthest reaches of the universe, the signs of something new and breathtaking. In September they would crank up the energy of his accelerator another notch—which was like increasing the power and resolution of that telescope—and they would focus in on whatever it was that they were seeing.

He expected that the answers would come by Christmas of 1984. The new physics would not go away, he said. Whether it was due to some misunderstanding on the part of his physicists as to what they thought the experiment was telling them, or whether it was due to some new physical process, would become clear in September. "Every physicist," Rubbia said, "is always concerned about the correctness of his assumptions until he has complete verification.

But as with the W and the Z, when you have complete verification, let me say, the thing somehow becomes a little dull. I mean, you have to live dangerously."

Again the theorists were taken by surprise. They had come to Bern, or to Washington, or to Harvard, where Rubbia gave the same presentation a week later, with all their talks prepared—six months of hard but not so exciting work about W and Z physics, or about their personal favorite theories, which until then had absolutely no experimental evidence to support them. Suddenly, everyone was talking about these new phenomena.

Now they just had to figure out what it all meant. They had a few leads. But the odd part was that no single theory seemed to be able to accommodate all the data. (The most difficult items to fit into any theory were the strange Z decays with the extra photons; but these had already begun to lose their credibility. Or, as one theorist put it, "The sex appeal with which they were presented has diminished.") So the theorists went about trying to find out what aspects of the data could be explained by their personal theories, and what aspects of the data they might consider ignoring.

Not that all theorists were so vigorously engaged. Theorists generally take one of two antithetical routes toward accomplishing their chosen tasks. Sheldon Glashow once called these the upward path and the downward path. The latter, as Glashow described it, starts from some brilliant idea, and then attempts "to go from a theory of everything to the mundane silly little effects seen at accelerators or on the earth." The upward path, as he put it, is a dirtier business— what Nobel Laureate Murray Gell-Mann refers to as ambulance chasing. It is the path that Glashow follows. "We listen to our experimental colleagues," he explained, "and try to glean little bits and pieces that do not fit into our standard and arrogant picture of the universe." These upward trekkers are known as phenomenologists, and they were the ones who were now getting agitated over the news from the collider. (The downward trekkers would prefer, in an ideal world, to ignore the experimentalists. They would like the chance to say, as Einstein was supposed to have said when he was asked what he would think if experiment eventually did not confirm his theory, "I would have thought that the Lord had missed a most marvelous opportunity.")

The phenomenologists now began to play their favored theories like lottery players who dream of winning numbers in their sleep. First there were those who believed in the sanctity of the standard model and who refused to accept that anything Rubbia or his colleagues were finding could drastically change the structure of their creation. On the pessimistic side of these were the ones who, because no single theory could explain all these bizarre results, preferred to write them off as extremely unlikely, but not impossible, decays of the W's and Z's. To them, the results had been wrongly interpreted by Rubbia as new physics, or were simply wrong. Among these was the prestigious theorist who told me that you would not find either him or his students working on such flimsy results, and that other phenomenologists "would turn around for any piece of turd in the street." Another theorist told me, "Some people who have real clout in the physics world made statements to me that Carlo had better confirm this or his ass is grass. He should put up or shut up."

On the optimistic side there were those, like Glashow, who tried to interpret the new physics with the most trivial possible alterations to the standard model—a fourth neutrino, a second Higgs boson . . . "It's one more epicycle" was how Glashow would put it later, after writing three papers on the subject that he admitted were all mutually contradictory. "The theory is pretty ugly to begin with. Instead of seventeen fundamental particles, now it's nineteen. Hotsy-totsy."

Then there were those theorists who liked the idea that quarks and electrons were not fundamental particles at all but were made up of even smaller fundamental particles. These people wrote papers speculating that the new physics could be indicative of this. After the strange Z decays, and the ensuing speculations that they could be the result of composite electrons, it seemed natural to try to explain the monojets and accompanying phenomena as a sign that quarks, too, had a substructure. When the sex appeal of the strange Z decays diminished, the theorists pushing these composite models grudgingly retrenched and decided they would wait for more data from the current run.

Alvaro de Rújula was one of this number. De Rújula, an aquiline-nosed, black-bearded Spaniard with the look of a slightly disheveled and cynical Don Quixote, was a Glashow protégé. He had spent several years with Glashow in the seventies working on theories to accommodate the high-y anomaly, and now he was leery of Rubbia's credibility. "One gets disgusted by the evolution of the things that they tell you," de Rújula told me, on the subject of the UA1 results. "One day they seem to tell you something, and the next day it's slightly different, and the day after that slightly different again."

Finally, there were the front-runners, who were working on the theory known as supersymmetry. Eight months earlier, a quartet of Spanish theorists from the University of Madrid—Maria José Herrero, Luis Ibáñez, Cayetano López, and Francisco Ynduráin—had proposed that Rubbia's collider should be able to produce certain strange particles, never before seen, such as photinos and squarks. In one out of a billion collisions, predicted the Spaniards, a squark would be created and would decay instantaneously into a normal quark and a photino. The photino would be undetectable, escaping like a neutrino through the detector and out into space; the quark would fragment into a jet. The result would be a monojet. When Rubbia reported at Bern that UA1 had seen monojets, it looked as though the Spaniards were holding the winning ticket.

The catch was that those photinos and squarks were creations of supersymmetry, SUSY for short, which was a theory in the purest sense of the word. It was an example of the downward path, and in the ten years during which it had come of age, experiments had turned up absolutely no evidence to support it. The theory had polarized the physics community. Many theorists found the mathematics of supersymmetry so beautiful that they doubted the Lord would miss such a marvelous opportunity. But others found the idea so extravagant that they simply refused to take it seriously.

The mathematics of supersymmetry demands a rather remarkable relationship between the stuff of matter in the universe and the stuff of forces. In the theory, the particles known as formions—such as quarks and electrons—which make up all matter, and bosons—such as photons, W's, and Z's—which transmit all forces, become somehow interchangeable. In a mathematical sense, at least, it would mean that matter and forces are equivalent, in the same way that Einstein proved that matter and energy are equivalent. It would also mean that, in terms of an ultimate theory of everything, at the instant of the birth of the universe all forces would have been encompassed in one superforce, and all particles, both fermions and bosons, would have existed as one species of superparticle. The idea was awe-inspiring.

The reality was a tad less impressive. Supersymmetry, like a fortune teller with delusions of grandeur, predicted the existence of a whole boatload of new particles. Not just one or two, but an undiscovered particle for every one already discovered. For every species of fermion, supersymmetry called for a new species that had all the characteristics of that fermion, but that was actually a boson. For every boson, supersymmetry called for a fermion. It was like matter and antimatter all over again, except that now it was particles and superparticles—or sparticles, in the lingo.

As though to accentuate the rift with reality, supersymmetrists gave their new particles even more fanciful names than usual: squarks, photinos, selectrons, smuons, gluinos, sneutrinos, staus, even higgsinos and gravitinos. Not one of these had ever had the grace to show up in an experiment, and not one of the known particles could be proven to be the sparticle of another known particle. It was not the best sign.

With a few exceptions, physicists smiled condescendingly at supersymmetry and followed other trails. The exceptions pressed on undeterred. In 1976, they struck upon a link between supersymmetry and gravity and created the theory known as supergravity. It was another pretty piece of mathematics, and a potential ultimate theory of everything, but it also seemed to be another step further removed from reality. Again, it was generally considered intriguing but irrelevant—a dazzling mathematical sleight-of-hand. It did not convince the cynics. "Einstein's general relativity does involve beautiful mathematics," Glashow once said, "but it was invented by Einstein for the purpose of doing physics. Newtonian physics forced Newton to invent calculus. It's not that some joker who invented calculus said, 'Here's some beautiful mathematics; let's make physics out of it.' That's not the way it works."

By the late seventies, the supersymmetrists were trying anything to bring their theory down to earth, even if it meant taking the long way home. They were building theories that existed in ghostly universes with as many as eleven dimensions, and then trying to reduce that number to four. With no success. Then in 1981, a series of papers, the first by Ed Witten of Princeton University, who physicists acknowledge to be one of the smartest men in the world, showed how supersymmetry could solve some of the hairier problems with the standard model. All it needed was for the yet unfound sparticles be only a little heavier than the W and the Z. Wishful thinking, cried many of the disbelievers. (Another of those papers was written by Howard Georgi, again a Glashow disciple, who later claimed he was embarrassed by it, the only problems it solved being trivial ones. Georgi has a photograph on the wall of his office at Harvard of a street sign with the warning: "Wino crossing." Beneath it is a man crawling across the road, a bottle of wine in his hand. The caption, penned in by hand, says: "The only evidence produced so far for supersymmetry.")

Others who had originally been skeptical, predominantly the Europeans, now considered converting. As Dimitri Nanopoulos, a CERN physicist, recalled, "I was visiting Rockefeller University at that time. At the moment I was leaving New York, I saw Ed Witten's paper in the library. I made a photocopy and read it on the plane —and I was screaming at everybody, to me it was so beautiful."

By 1983, European physicists such as the Madrid quartet were creating supersymmetry models in four dimensions that predicted that the desert of the standard model should be blooming with sparticles.

At CERN, where the supersymmetry contingent was strongest, they received the news of the UA1 monojets as though it were Gabriel trumpeting Judgment Day. After ten years they were confronted with vindication. They began to talk of the upcoming run of the accelerator in September as the SUSY Run, and about moving into the "S"—as in supersymmetry—sector. At the forefront was John Ellis, a self-styled prophet of the supersymmetry gospel. Ellis was a thirty-eight-year-old Englishman with shoulder-length hair, fading from red to gray, and a beard that fell across his chest and was holding the red better. Ellis was optimistic, he said; confident enough of SUSY to bet a Stanford theorist that they would have experimental evidence acceptable to Burt Richter, as judge, by the following July.

Ellis had been ignoring all the data that were not consistent with supersymmetry—just as he had ignored the high-y anomaly in the seventies. "If you want to interpret what's going on," he told me, "you have to exercise a lot of judgment, and you have to be prepared calmly to throw away pieces of the data that don't agree with your theory. If you try to fit absolutely everything, you're going to go crazy."

One of Ellis's chief disciples was Nanopoulos, a cherubic thirty-six-year-old Greek who had done his doctorate in England and his post-doc at Wisconsin and Harvard. Nanopoulos speaks English with a heavy accent, and has a tendency to leave his sentences dangling precariously in midair while he races off on some new tangent. He swore to me that as early as 1981 he was certain that supersymmetry was the true way, that they had only to prove it. What he could not understand was why the American theorists

wouldn't, in his words, play ball with supersymmetry. Not that he
was complaining. "That's fine with me," he said, "because now I
believe it's too late for them. The chips are down. If this is correct,
it is going to be spectacular. This is no joke."

Nanopoulos and Ellis were engaged in hard-core phenomenology: trying to take the collider data and fit to them the most
precise theory they could, and then from that theory predicting what
might be expected to show up in the future. And they had such a
prediction: If supersymmetry were correct, they said, then the next
thing to come out of the collider data when the machine was turned
on in September would be dijets—two jets of particles shooting out
of the collision, backed by nothing at all. It was an odd configuration
rooted in the theoretical decree that SUSY particles would have to
be formed in pairs, and therefore dijets—the product of the decay
of a couple of squarks or a couple of gluinos—should be as prevalent
as monojets. Ellis and Nanopoulos had been talking with the UA1
physicists, particularly Rubbia, about their predictions, and in response the experimentalists had concocted a new trigger for the
coming run that would be more likely to net dijets should they
appear in the collisions.

This arrangement had alienated some of the other theorists, particularly de Rújula, who thought that the supersymmetrists were
imposing their interpretations of the data on the world as though
they were cult leaders, professing to be unbiased as they forced
their own peculiar gospel down the throats of their flock. "What
they do," de Rújula explained: "First, they write papers in collaboration with the experimentalists, stressing strongly one of the interpretations and therefore injecting into the experimentalists an
inclination toward one interpretation rather than another. Convinced as they are of possessing the truth, they then go around
conferences giving only one possible interpretation and icing the
other as uninteresting."

De Rújula, of course, also had his bias. Coming from the Glashow
school and believing in the sanctity of the standard model, he considered his colleagues to be following a line of research that was
more closely akin to that of the medieval alchemists than to any
scientific discipline. "In spite of the fact that the alchemists never
found a single speck of gold in their remnants of frogs and bats, or
whatever they cooked in their moonshine," de Rújula told me, "they
kept on plugging at it forever. That was only based on faith, not on
fact. One has the impression that this is what is happening in super-

symmetry." As far as de Rújula was concerned, of course super-symmetry would fit the data from the collider, because it predicted the existence of so many particles that it could be made to fit any data.

With the theorists divided into ideological camps, the dijets of the supersymmetrists represented a sort of double or nothing bet. If they showed up in the 1984 running of the collider, the supersymmetrists could buy tickets to Stockholm. If not, they would all be set back to the starting line, and would have to begin scrambling for position on the next set of data. For the moment, however, all they could do was wait.

BOOK II

THE FIRST SUSY RUN

"Night is the time that graduate students are on shift, and they don't lie. They don't know how; they haven't learned yet."

—Leon Lederman,
director of Fermilab

9.

SEPTEMBER, 1984: PILOT SHOTS

"It's because of the way Rubbia is that it's so exciting. If we had a dull group leader . . . I mean Mozart was mad, wasn't he?"
—a British physicist working at UA1

The experiment sits in a hangar-sized building of corrugated metal, between hillocks on which sheep sometimes graze. The building is divided in half: one side houses row after row of generators that feed the magnets for the accelerator. The generators emit a perpetual background thrumming, but the physicists have been living with it for years, and they don't hear it.

The other half of the building is empty, except for an enormous crane suspended from a movable girder along the ceiling and, at the far end, a huge pit six stories deep. In the bottom of the pit are a few pieces of equipment, ladders, and tools. But the sides overflow with miles of cables as big around as fists, tied in bundles of hundreds, looping around the walls like some demonic electrician's idea of party decorations.

The experiment, which was constructed piece by piece in the pit, has been rolled on huge rails into the path of the Super Proton Synchrotron. The beams inside the accelerator will pass through the 2,000-ton bulk of the detector, and collide in its heart. Then that building-sized piece of electronics, the endpoint of all those cables, will churn out physics. Rubbia, with his usual hyperbole, calls his experiment "the most complicated piece of electronics in the world."

In mid-September, 1984, a few days after I arrived at CERN, a British physicist took me down to what is called the Mobile Electronics Carriage to give me an idea of the degree of complexity with which they work. The Mobile Electronics Carriage, or MEC, is three stories of electronics that look like white mobile homes piled atop each other, crisscrossed by blue catwalks. On the door of the second floor of the MEC is a sign that reads: "If you have nothing to do, don't do it here." Above a metal cabinet is another that reads: "Hands off. You will be shot."

Inside the MEC, the noise of cooling fans is reminiscent of being surrounded by a motorcycle gang in the Lincoln Tunnel in rush hour. The signals from the detector, from the umpteen thousand cables, hook into crate after crate of electronics that line the walls. The MEC is the transfer point for all the electronics in the detector. The input is the explosion of particles in the collisions, which the detector turns into electric pulses and feeds into the MEC. There they course through the millions of circuits that decide whether to feed them onto magnetic tapes, storing them for the perusal of the physicists, or throw them away and look at the next collision. "It's just a whole series of integrated circuits," the British physicist explained to me, "that make decisions on whether there's a lot of energy or not. It's like somebody threw a bucket of water over you. The equipment says, 'It's a big bucket; it feels pretty exciting.' Or, 'It's a small bucket; who's interested?' Just the same thing done over and over again."

Then he tells a story to indicate the mind-numbing level of complexity. "I went to push an innocent little reset button," he says. "And suddenly all the power went off: magnets went off, lights went off, fans went off, the beam went down. I thought, 'My God, what have I done?' It turned out that a technician elsewhere in the experiment was doing some soldering, and accidentally swung around with an iron bar and smacked the fire alarm and everything went down."

During the previous week the accelerator physicists at the SPS control just across the border in France had been slowly bringing their collider up to speed. They were using only protons in the machine, taking them in and whisking them through the four miles of vacuum-sealed tubes, then dumping them and taking another bunch. Their machine had not collided protons with antiprotons for fifteen months and they were adjusting the magnets that kept the beams in their orbits and the radio-frequency cavities that produced the electric fields that accelerate the beams. They were tuning their computerized timing so that when the beams were injected from the Proton Synchrotron into the collider, it would be done within the proper half of a nanosecond. In that time, light will travel only six inches. If it is off by more than a nanosecond, the beams will not

collide in the center of the experiments, but will do so in the darkness somewhere down the vacuum pipes. ("Could you imagine our
frustration," said the British physicist, "if we built this detector,
pushed it into position, and then the beams were always colliding
about ten meters down the road?")

All of this has been reset to control a new level of energy. From
270 GeV in each beam, the collider has been upped to 315 GeV. The
machine has also been revamped to achieve a higher luminosity
than ever before.

Luminosity is the brightness of the beams; integrated luminosity,
as it is called, is the big-money statistic to the physicist: the total
number of collisions generated between the particles in the two
beams. Total energy doesn't mean anything unless you get the particles to collide, and the more collisions, the more data, the more
physics. Integrated luminosity is quantified in the Jabberwockian
concept of inverse nanobarns. If you think of protons as targets in
a shooting gallery, and antiprotons as the bullets, then nanobarns
are a measure of the size of the target the protons present to the
antiprotons (barns as in "couldn't hit the side of . . ."). Inverse
nanobarns are the score of how many times you hit one with the
other.

In the first run in 1981, they scored 0.2 inverse nanobarns. In the
W run, they improved it to 28 inverse nanobarns, and in the Z run
to 153. This year, they were shooting for 500, which would mean
they would have more than three times as much data as ever before;
three times more Z's, W's, and the enigmatic creations of the new
physics that they expected this run to nail down.

For the moment, it seemed that the machine was off as often as it
was on. First it was politics: CERN was celebrating its thirtieth
anniversary, and Spain was rejoining the laboratory after quitting
thirteen years earlier. King Juan Carlos was in town with his wife
and two daughters, and Rubbia spent half a day with the royal
family, taking them through the experiment and into the tunnel of
the accelerator, along with a slew of the CERN management. (Meeting royalty and heads of state had become part of Rubbia's routine.
He had taken Pope John Paul on the grand tour, as well as Margaret
Thatcher, the Dalai Lama, François Mitterand, and quite a few others.)

A week later, the machine was shut down when the Kendrew
Committee came to CERN and was given the tour. The Kendrew

Committee had been established by the British government to reassess the value of their country's involvement in CERN, and was made up of distinguished British scientists. They had cocktails with some of the British contingent of UA1, and when a committee member asked one young British physicist whether he was worried that Britain would pull out of the laboratory, Rubbia was said to have answered, "Listen, we don't care what you do. Pull out. Stay in. Just get on with it so we can get back to doing physics."

Then it was lightning that shut the machine down. A bolt had struck the local power lines, the current had been tweaked—it only takes the smallest flicker—and the machine had kicked the beam out of its orbit. A thunderstorm anywhere within a 100-mile radius of the laboratory could drop the beam. In fact, virtually anything could drop the beam, from a fluctuation in a single cooling pump to an overzealous security guard testing one of the interlock doors into the beam tunnel, shaking it just enough to break the contact for a split second.

After the lightning hit, the accelerator physicists who control the collider agreed to leave the machine off, because CERN's showcase experiment, UA1, was not yet ready to run. UA1 had been down since July, 1983. In that time, they had added a new layer of detectors around the outside of the apparatus and had rejuvenated some aging electronics. When the UA1 physicists rolled their huge detector out of the pit and into position on the collider, it had been fine, but now it was leaking gas.

The UA1 central detector is filled with a mixture of argon, an inert gas that is perfectly safe, and ethane, which is very explosive. When the central detector is working, it has almost 30,000 volts sitting across its wires. Small amounts of oxygen leaking into that mixture of argon and ethane will destroy the detector's ability to track the path of subatomic debris. A large amount of oxygen could blow this wondrous experiment into tiny pieces.

After a day of crawling around the detector, the UA1 physicists located the leak, or at least one of the leaks. By evening the collider was running protons again, and the oxygen content in the central detector was down noticeably, although not yet to the normal level.

On September 23, a Sunday evening, I stopped by to see Rubbia in his office and ask him about the coming run. Rubbia's office is cluttered with papers piled about one foot deep on every available surface. There are a couple of floor-to-ceiling bookshelves exploding with black notebooks filled with more papers. There is a blackboard crammed with equations scribbled in chalk. On the walls are various mementos: the famous bagman poem by John Adams, a picture of Gersh Budker, an artist's sketch of the experiment, a few photographs of galaxies.

Rubbia is working on a paper, cutting and snipping and pasting. As he talks, he moves his hand to his mouth, holding it there. But he refuses a cigarette. Later, he chews on a piece of paper, a tiny rectangle, and then rips off another as the first one sits at the side of his mouth, forgotten. He looks nervous, and I ask him if that's the case. I was not yet accustomed to his notorious difficulty in sitting still. Most of his conversations with his physicists take place while he paces back and forth along the balcony outside the second-floor offices.

"Zero," he says to my question. "Absolutely not." He holds his hand out so I can see it's not shaking.

Frustrated, anxious?

"No," he says. "Absolutely not. You cannot do that. You have to understand that the people we have here are incredibly professional. It's like you go into an operation and you see someone with all their guts out, and you ask, do you feel nervous about cutting? If you feel nervous, you'd never be able to operate."

Rubbia has just flown in from the States, and he is playing with a pill case that is now empty of medication for sore throats. "The weather changes quite a bit," he says by way of explanation. "In the States, it was very hot." Thinking of Rubbia's flying, I remembered the British physicist who told me that he once flew with Rubbia to Rome. When Rubbia got settled in the jet, he leaned back, stretched comfortably, and said, "Ah, my first flight of the day."

Now Rubbia is telling me that it will be a couple more days before they get their antiprotons, and more than a week before they take any serious data. Once they start with collisions, they will have to carefully calibrate each piece of the apparatus, and verify that what they are seeing is real—not produced by glitches in the electronics.

Not that he isn't extraordinarily optimistic. Any questions raised by the new physics, the monojets and others, he tells me, should be answered by the end of the year, no matter how slow the start-up. "The choice you have on the monojet events," he says, "is what I liken to going to an extremely fancy French restaurant and asking yourself whether you want fish or steak. You are in the situation that no matter what you choose, you end up having a great time."

Then Rubbia delves into the fortune-telling business. He tells me that the monojets are most likely supersymmetry, and that the consequences are not just physics, but politics. The Americans are racing to build their Superconducting Super Collider, and the Fermilab people are all excited about turning on their Tevatron in a year. Now, he says, "the real physics, the gold mine, is right in our hands. Supersymmetry's one fundamental prediction tells you that new physics has the same value of energy and masses as the W mass. That will be the most productive and most exciting thing."

Finally I ask him what he thinks about the Zeta. The Zeta was the latest hot commodity in physics, a potential new particle that had hit the news in August. A group of California physicists working at DESY had reported finding it in the decay of upsilon particles. It had been in all the papers, and the situation had got a little out of hand, considering that nobody was too sure about the Zeta, especially the physicists who had discovered it. Some of the theorists, as usual, had ignored the pessimists and glommed onto the Zeta with passionate intensity. They had writen papers claiming that it was anything from a Higgs boson—not *the* Higgs boson, they said, but *a* Higgs boson—to a variety of different supersymmetric particles.

In August, when Rubbia heard of the Zeta, he immediately called a UA1 physics meeting. He had been worried: What if it was the Higgs? What if it was supersymmetry, and the physicists at DESY came up with the proof before UA1? Rubbia made copies of the conference report and gave one to each of his physicists. He considered digging back into the data and, as one UA1 physicist put it, "pulling that thing out come hell or high water." But Rubbia wasn't enamored of the evidence. "I'm not too convinced about this," he had told his physicists. "Maybe they left their signal generator on."

Now he says that, as far as he's concerned, Walter Sullivan of *The New York Times* wrapped it up when he called it "the mystery particle." This phrase came from the headline of Sullivan's article on the Zeta in the *Times.* "Eventually our destinies will meet," Rubbia tells me. "For the moment I don't see how. This mystery particle is a light particle, 10 GeV, and it could be something. Anyway, it may well be that this is a part of what we see."

I ask Rubbia if he feels any pressure over this. The supersymmetry predictions seem so vague that the Zeta could still end up in the

SUSY pantheon, in which case, the biggest discovery of the decade might have already popped out of someone else's machine.

"Look," he says, "I agree with you. You never let this hot potato cool down. You use it as long as it's warm. But we are so close to running. My guess is that if there is anything there exciting to be found, we'll probably have it in our hands before you leave. Our main problem will be to have it verified by others. You know the story of the other shoe? We got to have the other shoe."

And so it goes. The machine runs proton beams. The shakedown continues. During the day, the UA1 control room is packed. It is maybe 12 feet wide and 40 feet long, with over sixty television monitors. Some of the screens are black and white, some multicolored, and below various graphs and diagrams are blinking messages reading: "Waiting for an event . . . waiting for an event," or, "Waiting for a command . . . waiting for a command."

At any one time, a couple of dozen physicists might be in the control room, and they will all have a slightly harried air. They sit at their consoles, poring over computer printouts, modifying programs that define the criteria by which they accept an event as interesting or throw it away. They huddle around screens that indicate the state of some particular piece of electronics and confer quietly about what should be done, or who should be called to do it.

"You get that counter fixed?" one physicist says to another.

"Yeah," the other says. "You going to write it down in the logbook?"

"Probably not. It wasn't written in the book when it broke."

They always seem to be in a hurry. When they walk, they walk fast. When they talk, they talk fast. Like Alice's White Rabbit, everything is late, late, late. They are nervous. Nothing is working yet. The central detector cannot be turned on until the gas returns to normal, and without the CD, they have no physics. They have only eighty days of running, and they might miss the collisions that will lead to the next great breakthrough. Late, late, late.

Even at night, the control room is never empty. Usually, it's the trigger kids, two young British physicists and one American, who seem to be always there working. One night they are quietly dabbling with a part of the trigger system that decides whether the chambers that register the passage of muons have actually seen a muon energetic enough to care about.

At twelve-thirty in the morning, their work is interrupted. The phone rings in the control room and an Austrian physicist, the shift leader for the night—called the SLIMI, or shift leader in matters of information—picks it up, listens, and listens. It is fairly obvious to everyone in the control room that it is Rubbia demanding answers to questions. The Austrian replaces the phone after a few minutes. He says to no one in particular, "Bloody hell. We have no beam, we have no gas, and he wants to know what we're doing." Then he lights up a cigarette and paces across the floor. The others have seen this before and understand. A few shrug their shoulders; two of them laugh.

Now the wheels are turning. The collider is circulating pilot shots of protons while they work, and the controllers of the Super Proton Synchrotron are tentatively injecting beams of antiprotons for the first time.

Half the monitors in the control room reveal the same image: a screen full of numbers that are the vital statistics of the accelerator —the parameters of the beams, the magnets, the power supplies. A similar monitor is visible in every building at CERN. Few physicists can pass it by without stopping to see how many antiprotons are coursing through the pipes. At the bottom of the screen is a message: "The last one looked OK. The next one will be a triple pilot, which we will keep if it is good." A triple pilot is three bunches of protons at once, the number used in collisions.

A mechanical voice announces: "P-minus Check." It echoes through the empty building. On the screen, the words "P-minus Check" appear over the usual display of numbers and graphs. The collider is ready to inject antiprotons. A few minutes later, the loud-speaker announces: "SPS Ready," and the words appear on the screen. Now the collider is taking in protons every twenty-six seconds, flinging them around the machine, and dumping them. It is waiting for the antiprotons.

The mechanical voice says, "Cycle One." A few seconds later, it runs through a mechanical countdown from nine to zero. The trigger kids barely look up from their work. The television screens change not in the least. On one solitary oscilloscope, perched as though it is an afterthought amid all the high-powered ultramodern diagnostic displays, two parallel lines fluctuate. First the upper line contorts, indicating that the protons have been injected; then the lower line does the same for the antiprotons.

"Looks reasonably healthy," says one of the trigger crew.

"Well, it's better than what we've been getting," says the Austrian, glumly.

Every evening at six o'clock the collaboration meets in a conference room to discuss the running schedule and any problems they are encountering. There are nearly one hundred yellow chairs, placed in three even rows and along the walls. There is a blackboard in front, an artist's sketch of the detector, and the ever present television monitor mounted on the wall.

The physicists wander in for the meeting, and as in a schoolroom, they seem to first take the chairs in the rear or along the sides of the room. Those who choose the front row either have something important to say, or have arrived too late to get a seat in the back. (As one British physicist put it, "It's to keep as far from Carlo Rubbia as we can.")

When Rubbia is in town, the meetings run along a predictable course. Rubbia arrives from five to ten minutes late. One of his right-hand men, either Hans Hoffmann, the forty-one-year-old German in charge of UA1 technical matters, or Antoine Leveque, the co-spokesman of the experiment, describes the state of the accelerator and the plans of the SPS control crew. Rubbia will let these briefings go for a minute or two at a time before breaking in with a running commentary that is sometimes funny and sometimes brutal.

One day it goes like this:

"The main problem they had with the machine . . ." Leveque begins.

"Was UA1," Rubbia interrupts, and gets a few laughs.

" . . . was an excessive ripple in the power supplies," Leveque finishes. "And when they went looking for it, it disappeared . . ."

"I don't understand why those people can't find a ripple in the power supplies," says Rubbia. "There should be an easy way to do it."

"Tomorrow at two, they expect to have p-bars," Leveque continues, using the physics slang for antiprotons. (Antiprotons are written as p's, with a bar above.)

"What the hell were they doing the day before yesterday, the day before that, and the day before that?" Rubbia says. "As usual we have to sit back and relax."

Two days later it goes like this:

Rubbia walks in and from the door glances at the television screen.

"Something in the machine?" he asks. "P-bars? Shit, what are we doing here? Let's get to work."

And then Leveque gives his report. He is a small, shy Frenchman with whitish-gray hair. Rubbia gave him the co-spokesman job in 1983, when Alan Astbury left. In some ways, it was to pay Leveque back for being one of the first to support him on the collider project. Leveque is the straight man in Rubbia's vaudeville team. He wears a tweed flannel jacket that is a little too large, and he seems to take a beating frequently in these meetings. ("I don't get it," Leveque once told me over lunch, "Rubbia's certainly got some kind of aggressivity against me at the moment.")

The accumulator, Leveque explains at the meeting, had a problem with storing antiprotons. The machine people who run that particular accelerator want to dump three days' worth of antiprotons so that they can get in and fix the machine. That's three days' worth of physics.

"What do you mean, there's a problem with the stacking rate?" Rubbia says angrily. He's looking at the monitor. "Never saw so many p-bars in my life. They're rolling; they're rolling. Why do we have to wait? The machine is running, producing physics. I think we should work until the first time that we lose the beam. Look, we have to get going."

Hoffmann, who is one of the physicists most responsible for running the collaboration, breaks in to explain that the next morning they will turn off the machine for three hours so that they can change a broken pump in one of the cooling supplies.

"It means for us," Hoffmann says, "that for three hours tommorrow morning, we can't switch on the magnet."

"But there is no beam," Rubbia asks incredulously, "why do you want to switch on the magnet?"

Finally Leveque closes with the last words from the accelerator people in their control room across the French border.

"They're waiting for the experiment to get ready," he says.

"How can we get ready?" Rubbia exclaims. "They don't give us the fucking beam, how are we supposed to get ready?"

It continues. The SPS control room struggles to get their extraordinarily complex accelerator readied perfectly, so that they can give good beam to the experiments. The two collaborations try to get their extraordinarily complicated experiments readied so that when the beams come through, they will see the collisions. It's a symbiotic relationship that runs rough at this time of the year.

"You pick up the telephone to the SPS control room and say, where the hell's the bloody beam?" says one British physicist, telling his favorite story. And the guy in the control room, who he has known for years, says, "I don't know. Where do you think it is?"

By September 29, the central detector is working. The gas problems are gone. The physicists are finally watching collisions at 630 GeV. The collider is running three solid bunches of protons and antiprotons—what the physicists call dense shots, or big shots. On the largest screen in the control room the computer is drawing the tracks of particles, and of their high-speed annihilations, at about ten-second intervals. The tracks are red, blue, green, magenta, and yellow; they appear like a Monet representation of an explosion or a tidal wave. Some of the events are what they call beam splashes: a proton or an antiproton collides with some stray gas molecule in the pipes and splatters through the detector. In the good collisions it looks like the particles simply detonate at the center of the experiment, and all the debris is squashed toward the horizontal plane by some unseen force. Some slow-moving particles, trapped in the magnetic fields, spiral endlessly inward, leaving circular tracks of light on the screen.

When the CD is on and the machine is running, the displays that previously read: "Waiting for an event," or, "Waiting for a command," are no longer waiting. They are now lit, blinking in magenta letters "CPU was running phase 1," or, "spy any event". The atmosphere in the control room has changed. The harried feel is gone, replaced by one of relief. The machines are running. There are no more shadowy doubts; only line-tuning remains. They will have physics.

On Sunday, September 30, Rubbia strides into the six o'clock meeting looking, as he often does, ready to pick a fight. Once the meeting begins, however, he calms down. Hoffmann opens. The SPS is running, he says. There will be a dense shot soon.

Then Rubbia takes over. "Okay," he says. "Cross your fingers. Chances are we will get a good piece of luminosity. The machine seems to be well behaved."

Rubbia demands status reports. If his physicists run on for more than a minute or so, he interrupts them. If they say only a few words, he demands more information. If somebody repeats old news, Rubbia's response is "Yeah, we know that."

To the trigger crew he says, "The question is, are you personally confident that we have a working trigger?" The answer is to his liking. He moves to the central detector and is told that the gas is normal, the voltage 100 percent. The apparatus is running like a dream. He smiles and moves on through the various layers of the apparatus. When one physicist tries to speak at Rubbia's fast clip, Rubbia interrupts and asks him to slow down, "so we can understand what you're saying." When another gives his report in French, Rubbia fires back his questions also in French.

When Rubbia is not talking, he is pacing about and looking at the television screen. It's not at all unusual for him to turn his back on someone and begin his pacing as soon as he has given that person the floor.

"Now," Rubbia says to the sixty or so physicists in the room, "I would like to propose a strategy to see if you agree." He searches for a piece of chalk and writes "Strategy" on the black board. Then he says, "One. Try to get whatever we had last year working and try to run with it for two nights—tonight and tomorrow.

"Suggestions?" he asks. "This was a statement for discussion." No one answers.

"Two," he continues, "use two days' shutdown for feedback. And three, Thursday, when we come back, get the extra stuff in the system going."

Then Rubbia repeats it in more detail. When a German physicist timidly admits that they have already hooked up a new untested trigger system for the muon chambers, Rubbia berates him for not following strategy point number one, which is now law. The rest of the physicists stare glumly at their feet or look uneasily at each other, waiting for the tirade to pass. ("Quite often," a British physicist explained later, "what Carlo says in these meetings is the essential thing to say. It's just that I don't treat people that way.")

Finally Rubbia wraps up the meeting. "Is everybody satisfied by the current state of affairs?" he asks. "Anybody want to make any statement, suggestion, remark, criticism?"

The collaboration remains quiet.

To the German he was abusing five minutes earlier, he says, "You have plenty of stuff to chew on. Don't try to die from indigestion." He smiles, and the German smiles.

"Look, folks," Rubbia concludes. "Let's get going. The beam is there. We have a run."

10.

OCTOBER, 1984: THE PRINCE

"Hey, this is the most exciting fucking experiment in the world. This is it. If you're not here, you're dead."

—James Rohlf, UA1 physicist; said at a Geneva hang-out known as the Boot Bar

No one at UA1 doubted that Rubbia would get the Nobel Prize; they only wondered whether the day the prize was announced —which would be October 17—would correspond with one of the infrequent days that Rubbia would be at CERN. Rubbia had said that he would be in Texas until the 16th, and would go to Trieste to lecture on the big day. When van der Meer was informed that Rubbia was planning to be out of town, he laughed. "I think that's disgusting," he said. "He's leaving me alone to face it all." And then van der Meer wondered if maybe Rubbia knew something he did not. "Maybe nothing happens," he said, pensively.

Rubbia arrived early on the morning of the 17th. He wore his business suit, had his bags, his trenchcoat, and his smile. He worked on his physics and waited for the phone to ring. He paced the corridor in front of his office and smoked Gauloises bummed from his secretary. He watched the monitor that charted the condition of the collider. And he left at ten o'clock for Trieste, where he planned to give a lecture with Abdus Salam. "Of course," Rubbia related, "Salam said I should refuse, because I might get the famous phone call and should stay by the phone. I said I have to live my life." As it turned out, there was an air-traffic strike in Italy that day. Rubbia found himself sitting in a cab on the highway outside Milan when the cabby's pop-music station announced that Italy now had its first Nobel Laureate in physics since Enrico Fermi.

When I asked Rubbia the next day what would happen now, he told me: "Right now, I'm going to consume a considerably large amount of alcohol in the form of champagne, which is a mixture of alcohol and carbon dioxide, and then I think we're going to continue."

That afternoon, the Harvard contingent of Assistant Professor James Rohlf, two graduate students, and Mike Levi, a CERN fellow, put three bottles of Dom Pérignon on ice in red plastic buckets and waited for Rubbia to return from lunch. When he did, Rohlf dragged him away from a Swiss television crew and popped the champagne. They toasted Rubbia with the cameras rolling, and Rubbia himself devoting most of his attention to a pretty blond reporter from Zurich. When Levi wished him good luck on his next prize, Rubbia responded, "Which one, the second or the third?"

It was at the next party, in the cavernous building of the experiment, that the news broke that the Zeta particle had apparently been a mistake. Inside were two long tables laden with food and drink, and a couple of smoking barbecues. Rubbia showed up with van der Meer at 4:15 P.M. For once, no one was working in the control room, or anywhere else, for that matter. After a few toasts, Rubbia and van der Meer climbed on an elevated platform and Rubbia gave his thank-you speech.

"You folks did all the work," he told them, "and I am getting all the money. Wait, sorry. He is getting the other half." He pointed at van der Meer. "Now he is going to tell you why he deserves half the money."

Van der Meer said modestly, "Thank you very much for finding those particles, without which I wouldn't be standing here today." Then the drinking continued as the smoke from the barbecues got thicker and thicker.

Amidst all this self-congratulation, a German physicist, recently back from DESY, described the demise of the Zeta particle to a few of the young American physicists. He said simply that the new results that had come from DESY had shown no sign of the mystery particle. It was gone—vanished into the swamp of background like the Loch Ness monster after another tentative siting. No one knew yet whether the original signal had been a misunderstanding of their detector and its capabilities, whether it had been created subconsciously by the physicists, or whether it was something more complicated. But it was gone. That's all they knew.

The parties ended the next day with one huge bash for the entire laboratory. Rubbia had more parties to come in Italy, where he was on his way to becoming a national hero, and one at Harvard, where he told his Ivy League colleagues, "You people think it's hard to win a Nobel Prize, but it's easy. Trivial. Just put protons and antiprotons in a box, and shake them up, and then collect your prize."

Between the discovery of the Z and Rubbia's subsequent Nobel Prize, the UA1 collaboration had changed. Most obvious was that they now had the aura of success. It was as though the whole collaboration had a collective swagger. They had proven themselves. It was as simple as that. They had no more worries, no more fears; even their sense of competition had lessened. They talked of how UA2 had lost, and how they had firmly established that their own detector was far and away the superior creation. They had the monojets as their next breakthrough, the hottest thing in physics. The fact that the UA2 physicists believed that they, too, were onto a breakthrough, that they too had perhaps seen the signs of new particles, did not faze them. UA2 must be wrong, they said, as the Zeta had apparently been wrong. The world would now turn only to UA1 for the important physics.

Less obvious was that many of the key individuals in the UA1 collaboration had slipped away. Sadoulet was one of those who had left. He went off to Berkeley on a one-year leave of absence to work in astrophysics. No one expected him back. He was tired of playing devil's advocate. "Your enthusiasm to fight decreases," he said simply. Calvetti had left, breaking off bitterly with Rubbia about his responsibilities now that the central detector had been built. To Calvetti, both he and Sadoulet were too independent to last under Rubbia's control. "In UA1," he said, "you can be brilliant, but it is all the time reflected light. It is not your own light. As soon as you try to have your own light—just a little light—then you are pushed out." Astbury had moved to Canada and had been replaced as co-spokesman by Leveque. Michel Spiro had quit after criticizing the newness of the "new physics" in the monojet paper. Denny Linglin, a Frenchman who—with Rubbia, Sadoulet, and John Dowell —was one of the original four to present the UA1 proposal to CERN in 1978, had left to become head of the physics department at Annecy.

The exodus was partly a natural regeneration, and partly a consequence of Rubbia's style. It was explained to me by Kate Morgan, a young technician who had been with the experiment since 1978. "There was a feeling in the group to find the W," she said, "and when we found the W, and even when we found the Z, that group feeling was very high. And what happened since was we didn't really have any reason to stay as a group. We had all come of age under Carlo. Okay. We learned our lessons, we took our knocks, and in some ways we had gained a certain expertise. We were good and we knew it. We had been vindicated in the eyes of the world; but we would never be vindicated in Carlo Rubbia's eyes."

With the departure of all these people came a change in the group structure. The physicists who had left had not only been responsible for building much of the apparatus, they had constituted a second level of command that could to some extent control the analysis. This meant having the strength, the wisdom, and the seniority to fight Rubbia during those times when he went overboard. Now they were gone, and the vacuum had been filled by a new generation, what some of the British physicists called "Carlo's Disciples." It was also these British physicists who divided the collaboration into three factions: us, them, and the rest.

"Us" were the physicists who did their jobs as well as they could, had as little contact with Rubbia as possible, and preferred it that way. These were mostly regulars who spent much of the run around CERN, babysitting the more delicate equipment, or working on new technologies to add to the detector in the future. Most of them rarely worked on the actual physics, but they were indispensable for the hardware, and in fact they had built much of the apparatus and knew it blindfolded. If the machine broke down, they went into the pit to repair the damage. They had all been with the experiment from its inception, and many had survived, they would always say, because they did not work for CERN and therefore, theoretically, Rubbia had no power over them. These were the people Rubbia most needed; they made up the bulk of the experiment, and in a way they were irreplaceable.

"The rest" were those who arrived after the experiment was launched and most of the hardware work completed. They were post-docs and grad students and visiting fellows. They had no critical responsibilities because no particular piece of the apparatus depended on them. They seemed to come and go with each new run, and it was hard to keep track of the new faces. Occasionally one of "the rest" would become an "us" or a "them," but for the most part they worked on minor chunks of analysis and remained relatively anonymous.

"Them" were Carlo's Disciples, a small core of young physicists who worked closely with Rubbia (too closely, according to the "us" contingent). These people took or were assigned the most critical tasks in the analysis. They were all very bright and very ambitious. Some of them, like Jim Rohlf, came from Harvard with Rubbia. Some of them, like Jean-Pierre Revol, worked for other institutions—in his case, MIT—and fought Rubbia's constant offers and pressure to switch to CERN or Harvard, where he would be very much in Rubbia's shadow. Some of them, like Felicitas Pauss, an Austrian, or Steve Geer, an Englishman, had taken up such offers, and now worked directly for Rubbia.

These disciples had no twenty-four-hour machine responsibilities—for the most part because they had come after the experiment had been built—and no family responsibilities. They were young and aggressive. They put in the longest hours on the experiment, but did so on the physics analysis, not the apparatus. Rubbia loved it, and expected it. "He tends to think," Leveque had told me, "that anybody who is not living a hundred percent for UA1 is not worth having in UA1." Even after working their eight-hour shifts in the control room, they would then move to the analysis for the next eight or ten hours. Only then would they go home to catch a few short hours of sleep. They did not have teaching responsibilities during the run, so they did not have to fly home as soon as their shifts ended. They had jumped into the analysis in the first place because it was where they could have the most impact. They then tended to monopolize it because they were able to work more hours than the others and because they were willing to work in a close proximity to Rubbia that the others found either too dangerous or distasteful. They played physics all out, the way Rubbia did. And they exhibited, to a lesser degree, his relentless greed for discovery.

Although Rubbia's disciples were not all Americans, even those who were not somehow seemed to be. Traditionally, Americans had a more obsessive attitude toward physics. They seemed perfectly content to work all the time, day and night—particularly night—and showed no apparent desire to sleep, to talk to their families, take vacations, or live outside the laboratory. UA1's computers were free and ran fastest at night, and much of the grunt-work of the analysis could be done painlessly.

The Europeans on the other hand were mostly daytime workers, "except for a few crazy ones," as Jim Rohlf put it. They seemed to want to enjoy their lives as much as unlock the ultimate secrets of the universe. The Americans, and in fact all of Carlo's Disciples, made no such differentiation—the two were irrevocably linked.

"The directors of CERN always say they like to have Americans there," explained Bernie Hildebrand of the Department of Energy, "because they're good examples of dedication and hard work. Carlo may not be an American, but he acts like one."

Unlike the CERN management, many of the Europeans on UA1 did not appreciate the influx of Americans, or the way Rubbia seemed to grant them special privileges. They were cashing in without paying their dues. Where were they when the experiment was being built? As Calvetti complained, "There are many people who are now talking at conferences, presenting papers, and so on, who came at the last day of the last run. This is not acceptable." Or, as Eric Eisenhandler said, admitting that although he was born in the Bronx, he sides with the Europeans, "At the beginning, it was always Carlo but not Harvard, Cline but not Wisconsin. Suddenly these guys are appearing. One has slightly funny feelings about anyone who joins when things are going well."

But for Rubbia, these young physicists represented more than just brains; they represented political leverage. For instance, Rubbia took Rohlf to meet the Department of Energy officials in Washington within six months of the time he joined the experiment. Anne Kernan, who had been the sole American leading an official collaborating institution (University of California at Riverside) at UA1, had until then enjoyed free access to the bureaucrats at the DOE, and the power and funding that came with it. Rubbia had used Kernan's Riverside group to funnel American money into UA1, thus relieving him of the necessity of getting money from Harvard.

Now Rubbia needed to convince the DOE that he had a strong Harvard contingent at the experiment, a lot of bright young kids who were cleaning up on the European accelerator. Rubbia wanted more American money in order to pay for the upgrading of the detector for the coming runs, and for that he needed people like Rohlf and Steve Geer. The DOE bureaucrats were tickled by the fact that bright young American physicists—or at least Rubbia's adopted Americans—were taking over the experiment. They thought of it as the American way. "Americans not only work hard on hardware," Hildebrand told me with patriotic fervor, "but they work ultra-hard on the analysis. And when they do—this would insult a lot of people —they tend to take over the experiment. We're in there punching hard, getting the product out. I'm talking like a businessman, but it is a business; the business to get physics. The fact that Americans are strong in analysis is great, and we try to fund the experiments so they will be."

The natural abode of Rubbia's disciples any time between four in the afternoon and four in the morning is the Megatek room. It is located on the ground floor of a two-story building that is simply two corridors one atop the other, strung between two other identical buildings that go off at odd angles. The building's only discernible characteristic, setting it off from a couple of hundred others exactly like it at the lab, is that it is close to the main administration building and therefore to the coffee and beer in the cantina.

Inside the Megatek room are scattered half a dozen computer terminals. One wall is lined with shelves and black notebooks with incomprehensible alpha-numeric titles; another has a bulletin board with photos and printouts of famous UA1 W and Z events thumbtacked to it. In the center of the room sit two large pieces of technology that look as though they were made for the ultimate video game; these are the Megateks.

The Megatek is a computer system that displays the collisions of protons and antiprotons in three dimensions. With a hand control —known as a mouse in computer jargon—a joystick, and a keyboard, a physicist can take a computer-enhanced three-dimensional reconstruction of a collision, and the subsequent decays of the particles that are created, and rotate it, enlarge it, shrink it, or even turn it upside down. Rubbia wanted to get a hold of the Megatek as soon as he heard that some hyper-imaginative computer wizard had cooked up such a thing. Not only could it create instantaneously what the best graphic device would struggle over interminably, it could do so gloriously. Rubbia claimed that the Megateks were the very same devices that science-fiction film makers used to produce their space battles in films like *Star Wars,* and that the computing power UA1 used was no less than that used by Hollywood.

The Megateks are at the heart of Rubbia's physics. Those physicists in the collaboration who had worked with bubble chambers argued that in the earlier runs, Rubbia had given too much weight to the Megatek as an analysis tool. Rubbia and his disciples, they said, had been deluded about the value of studying the beautiful three-dimensional pictures of the collisions on the Megatek. But, of course, it was Rubbia's show.

"You have all this computing," he told me, "but the purpose of all this tremendous data analysis, the one fundamental bottom line, is to be able to let the human being give the final answer. It's James Rohlf looking at the fucking event who will decide whether this is a Z or not."

Rohlf had become the resident Megatek wizard. It seemed sometimes that he could look at an event, and with a sleepy expression pass his verdict within five seconds and never be wrong. He would sit there in his tweed jacket and khaki slacks, slouching forward over the mouse, staring at the events, and if he liked them, he would always say something like, "Nice, huh?" That was all. Nice, huh?

Rohlf was a little heavy, with a frumpy demeanor like that of a grown-up Huckleberry Finn; he had a sort of cocksure honesty. He liked to stand with his eyes half-closed and his chin jutting out a little as though he were daring you to take a poke at it. He tended to be scatalogically outspoken, but somehow he got away with it. It might have been because he was Rubbia's number one disciple. It might have been because he was very bright, and he knew it, and much of the physics community knew it. Rohlf was a firm believer that if you were bright enough, you could pull off just about anything.

Rohlf had grown up in Minnesota and spent his high school years cruising in '57 Chevys, playing the trombone, drinking beer, and chasing girls. He was the kind of kid who might have been the smartest in his high school class, but you'd never know it because he was having too much fun. He floated through the University of Minnesota, studying physics and math without taking either too seriously. Then he got to graduate school and, as he said, "I realized that the physics was not the shit that was in the books. The physics was really with the people who were measuring the particles. That was what was the challenge."

He went to grad school at UCLA, and worked on an experiment at a Berkeley cyclotron. He moved on to Fermilab and Caltech when he decided that he might be brighter, or at least more driven, than anyone he was working with. "After a while," he told me, "I realized that these guys were turkeys, and that I had to get in someplace serious."

Turkeys? I asked him. Not as smart as you were?

"It's hard to tell how smart somebody is," Rohlf said. "Let's say they weren't as dedicated. Probably they weren't as smart. These guys didn't really understand how to get at the physics. They had lots of fun, but, for instance, they didn't put the proper effort into the computing, and they didn't put the proper effort into electronics. They were sort of dabbling around."

After getting his Ph.D., Rohlf spent two years on a Harvard experiment located at the Cornell Electron Synchrotron, and then he decided that UA1 sounded more interesting.

"I arranged to have dinner with Carlo," he explained. "My first real meeting with the great Carlo Rubbia. We spent three hours having dinner and we drank a bunch of wine. He was telling me all the great things you could do with the experiment. At the end, he said, 'You know what, I don't need you on this experiment. I have a hundred and fifty physicists at my experiment. I don't need you,' and he got in a cab and drove away and I went back to Cornell. Then I find out afterward that he got back to CERN and told Alan Astbury that he had met this tremendous young physicist, very enthusiastic, who is coming to work for us immediately."

Rohlf told his boss at Cornell that if he didn't like UA1, he'd come back to Cornell. But, as he said, "After I was here a week, I knew I'd never go back."

He started in June of 1982, and "got the shittiest jobs to do, which is understandable." But by the end of the year, in the middle of the W hunt, he had ensconced himself in Rubbia's heart. The UA1 physicists had discovered maybe two good W candidates by the time the laboratory closed for the Christmas break. Most of the physicists immediately went home. Rohlf stayed and scanned the remaining data from the run for sixteen straight hours. "Then," he said, "I left a message for Carlo that I'd scanned through all the data and there were three more W's, and you don't have to worry about it. When Carlo got back he said, 'It looks really nice,' and he was smiling. That's when I really felt I had gained enough knowledge to be able to set a discriminating line: This is shit below it, and this really is something that is going on above it."

Like all of Rubbia's lieutenants, Rohlf tended to do things to extremes. He bought a beautiful white used Mercedes on his very first day in Geneva, because he had promised himself he would do so as soon as he got to Europe. At the last count, he owned more than half a dozen different recordings of Beethoven's Ninth Symphony.

When Rohlf wasn't working, he liked to take his colleagues and graduate students on drinking tours of Geneva, which were often disappointing, as Geneva is a conservative city. ("The most depressing Monday morning I ever spent," said one of his drinking partners, stealing from Mark Twain, "was a Saturday night in Geneva.") They often went in Rohlf's Mercedes, beginning at some local auberge over a great country meal and a liter bottle of local red wine, or at Wendy's over a burger and beer, and then taking off from there. When Rohlf hooked up with Dave Cline—the two had become good friends—these tours would last most of the night, and throughout them you would be treated to Cline's encyclopedic tale-telling and Rohlf's succinct and unambiguous assessments of everything and anything. Often they ended in the early morning hours at a Genevoise biker hang-out known as the Boot Bar, where black-leather-clad customers order beer in enormous glass boots. The conversation during these evenings rarely veered from physics and physicists. The range of topics included Sam Ting—a favorite of high-energy physicists—the lack of sex in physics, the relative merits of UA1 and UA2 and the physicists thereof, and Rubbia's latest episode of politicking.

These grand tours would eventually end back at the laboratory, where the physicists would pull some chairs up to the Megatek and do a little scanning before going to bed. Unlike the Genevoise bars, the Megatek stayed open round the clock.

One of the physicists often to be found in this group was the Englishman Steve Geer. Geer was turning thirty, tall and slender, with reddish-blond hair, a narrow face, and a reticent manner. He had come from an exemplary college career at Liverpool, and had started at UA1 while still completing his thesis on another CERN experiment. "I was analyzing my data and writing my thesis past midnight," he told me, "and between nine in the morning and midnight I was working for Carlo." When he finished his thesis, he did some hardware work for Rubbia, and put together his own independent analysis on some of the early physics.

Then he turned the Megatek room into his own personal fiefdom. He had been given the responsibility for organizing the scanning for jets, after UA2 had announced that they had them in Paris. "That brought me into a controlling position as far as the Megatek was concerned," Geer explained. "When we had the W run, my life, at the time, was to spend essentially nineteen or twenty hours a day in the Megatek room. I made sure that every single damned event was understood and digested. I shouted and screamed at the people designing the selection program until it was reasonable, and then I processed the final output of the Megatek. I was completely in control of that data coming out. There wasn't a single exciting event that could come out of UA1 that didn't go through my hands first. It was worth the twenty hours a day to be in that position."

On the night of October 3, Rohlf sat with Geer at the Megatek, Rohlf in his usual khakis and tweed, Geer in a sweater and jeans. They had fed the latest data tapes into the computers, some of the first data from the run. They had gone out for Cokes at the cantina and were now waiting for the computer to dish up the first event. An event, or collision, appears as a colorful explosion occurring within a schematic diagram of the detector. At the center of the screen is a cylinder for the central detector, surrounded by a larger cylinder for the gondolas—the electromagnetic calorimeters, which catch those particles with electromagnetic energy—and then a rectangle for the C's—the hadronic calorimeters that catch hadrons, particles that are made of quarks.

The first event, to a neophyte like myself, was a mess. It had thirty to forty particle tracks of varying colors—the computer indicates the amount of energy in each track with a different tint—some arrow-straight, some curving gracefully out of the collision. One arrow was boldly cross-hatched: the computer had added up all the energy in the collision, decided that somewhere it had missed a particle—maybe a neutrino, maybe something that had snuck through the detector—and had drawn this arrow to show which way it went and how much energy it carried away.

Rohlf punched a button on his keyboard, and all the tracks in the event disappeared, save for a dozen of the highest energy, one very bright track, and the missing energy arrow.

"The energy was 83 GeV," said Geer, looking at a computer printout of the critical numbers of the events they were about to scan. "Take a picture of this."

"I did," said Rohlf.

Rohlf slouched in front of the screen, working a mouse with his left hand and a joystick with his right, and taking occasional sips of his Coke. He punched a few keys on the keyboard and the computer informed the printer to make a copy of the image on the screen.

When Rohlf moved the joystick, the computer rotated the image. The collision and the tracks of all the debris appeared to be frozen inside a palm-sized lucite cube, and it was as though Rohlf was holding it in his hand and turning it this way and that. A cosmic paperweight. When he zoomed in on the collision, it was as though the universe in the screen image was the real one; yours had disappeared. Suddenly you found yourself in an Alice in Wonderland world where dimensions were a trillion trillion times smaller.

"I'm trying to figure out if the shower is caused by a single track or not," Rohlf said, peering at one particular corner of the gondolas. When a particle hits the calorimeters, it creates a shower of secondary particles that are then detected by special light amplifiers called photo-multiplier tubes. Inferring the original particle from which the shower is created is like trying to draw a portrait of a man when you know only the combined characteristics of all his great-grandchildren.

"There's at least one jet in the event," said Geer, scribbling a few numbers in his notebook.

"I think it's a W," Rohlf said.

"Looks very much like a W to me."

Rohlf had changed the image on the screen so they were now looking at the path of the brightest track through the central detector. When the particle streaked through the CD, it ripped electrons off the gasses, which then drifted over in the electric field until they intersected the wires of the chamber. The screen showed the position of the "hits" on the wire, and the computer had reconstructed those hits as a single straight track of the particle.

"It looks like a genuine electron. I think it's a W," Rohlf said.

He tapped a few keys on his console and the screen changed again, back to the entire detector, with just the two tracks: the straight line of the electron, and the cross-hatched missing-energy arrow going off to the side. Rohlf and Geer ran through the cuts one by one. If this is a W, how can we prove it? they asked of each one. What could be wrong with it? How could we be fooled? What could make us think this is a W?

"Okay," Rohlf said finally. "It's a W plus jet." It was the collaboration's second W of the year.

"Now can we make a W list?"

Geer smiled. Last year's discovery was this year's confirmation that the experiment was rolling again, that the universe had not changed noticeably since last time. It was a rhetorical thought, but a comforting one.

They moved onto the next event. The laboratory was as quiet as it ever got, with only the whirring of fans, keeping the computers cool, in the background.

On the screen was the schematic of the central detector, but no collision. Instead, one row of squares, representing the cells in the gondolas, were all brightly lit. The cells had misfired and faked a collision with huge missing energy, and the detector had taken the event and put it on tape.

When the physicists had reprogrammed the detector and put in the new trigger that would fire on any event with large missing energy, they had hoped to net more of the monojets—and maybe the dijets that the supersymmetrists had predicted. But the cost of this was that any energetic particle that escaped unrecorded into a crack in the detector—and the detector was not exactly hermetic— or any collision between a particle in the beam and a stray gas particle, which might spray one side of the detector with particles and not the other, would be saved on tape. Last night, the electronics in the gondolas had misfired for five minutes, long enough for forty junk events to appear on tape.

Rohlf and Geer looked through the events as fast as the computer could display them, which was about one every ten seconds or so. One event was a misfire in the gondola; the next was a beam-gas collision; and the next was what they call a hedgehog, a collision that was so inpenetrable that they would never make sense of it. It could be chock full of bizarre undiscovered particles, but they'd never be able to tell. Rohlf hit the kill button, sending the image off into silicon obscurity. It reminded me of fishing, throwing a fish back after it was caught because it was inedible.

After another half an hour, they came on event number 455.

Rohlf 's reaction was instantaneous: "Fuuuuck! This is a really nice event. This is the type of event we have been looking for."

The collision had spewed out two massive jets, and the computer had added an enormous missing-energy arrow, so big it looked like it could nail your feet to the floor. The rest of the collision was clean, no more than a handful of stray tracks.

"The whole event has perfect alignment," said Rohlf.

"It's super," Geer affirmed.

"We've got to get the jet mass," Rohlf said, and fiddled with the controls for a few minutes. "It's 54 GeV. The total mass is 109. This is a very nice missing-energy event. It has two very clean big jets, a smaller third jet, and large missing energy."

They were peering at a denizen of the universe that may never have been captured before. Ever. It seemed exactly what super-symmetrists like Ellis and Nanopoulos had been predicting. But

Geer and Rohlf were both tired. It was late. How excited could they get?

Rohlf leaned back, massaging the back of his neck with his left hand, scratching, yawning. Geer sat up, writing in the logbook as one bad event after another flashed by. Sometimes the screen elicited glimmers of interest before closer inspection revealed some machine screw-up. Sometimes it was a cosmic-ray particle that had been caught traversing the detector and faking the apparatus into thinking that it had captured something more exciting.

Rohlf's fingers stayed on the mouse, the cursor stayed on the kill position. The finger moved, click, click. It was quarter of one.

The last event of the night was momentarily interesting, and then turned out to be nothing but a cosmic-ray particle.

"Okay," Geer wrapped it up. "We've got one superb missing energy, one W. Once we debug the system, then I bring in a thousand scanners to do all the work."

Rohlf ran through a quick calculation in his head. Was this a good night or an average one? "We should get more than one W a night," he said. "One Z a week, two if we count the muons. For the missing energy, we know from the rate last year that we get five of the type with jets, two of the type with photons. So that means these should come about one a week or so. All this of course depends on how well the machine performs."

Geer took a copy of the two-jet missing-energy event. "Give a copy to Carlo?" he suggested.

"He'll throw it in the garbage can," said Rohlf.

"Not the two-jet missing energy," Geer said. "He'll go nuts."

"He'll call a press conference," said Rohlf, "or he'll throw it in the garbage can."

"Or he'll do both."

Throughout October, it sometimes seemed as though the last thing people were talking about was physics. If it wasn't the Nobel prizes, it was the UA6 situation that had everyone talking. Underground Area Six was a small experiment run by two Swiss physicists, Claude Joseph and Louis Dick, and Rod Cool, an American from Rockefeller University. The experiment was set up under a little metal shack out in front of UA1 and looked a bit from ground level like a surplus electric generator that somebody had plunked down and forgotten about. It was designed to spray microdroplets of hydrogen into the path of the proton and antiproton beams, and then watch what happened when the particles in the beam scattered off the hydrogen. It was an interesting concept; probably not Nobel caliber, but good bread-and-butter physics.

What made UA6 controversial was that it was blowing its hydrogen gas into the vacuum pipe of the accelerator 100 yards down the beam line from Rubbia's UA1. This didn't sit well with Rubbia. "We are spending millions of francs in order to clean up the vacuum chamber," he told me one day, while driving back from a six o'clock meeting, "and the vacuum is as good as the vacuum on the moon, and those guys over there start blowing gas on it."

Since the machine was four miles in circumference, there were a lot of rumors about why UA6 had been put so close to UA1. The most likely was that CERN did not have the money to spend on the facilities it would need for another underground area—a parking lot, a guard at the gate, etc.—so instead put it within UA1's area. (Rubbia had his own ideas about this. The director of research when UA6 was approved was Erwin Gabathuler, one of the few men in the CERN management who consistently stood up to Rubbia, and Rubbia liked to refer to the experiment as "Gabathuler's Revenge.")

The physicists of UA6 had been expected to test the degree to which their gas jet would disturb UA1, but for one reason or another the tests were never performed—usually because the lab was too busy with Rubbia's collider to allow it.

Now Rubbia didn't want tests, he wanted UA6 shut down. He wanted them to go play somewhere else. The UA6 physicists claimed that once they made the tests, Rubbia would find that they did no damage. Rubbia didn't want to take any chances. He was looking for physics effects that were so subtle they manifested themselves in one out of every billion collisions. He would not take the chance that at some future physics conference somebody would raise his hand and say, "Gee, Carlo, you say you've discovered this new particle, but how do we know it wasn't simply UA6 spitting hydrogen atoms into your detector?"

Rubbia instructed his physicists to shut down immediately if UA6 tried to test. "If somebody calls you up at three in the morning, and says"—Rubbia imitates a little girl's high-pitched voice—" 'Can I please make a test?' just remember the story of the *Petit Chaperon rouge* and the Wolf."

On October 4, during one of the daily meetings between the physicists running the accelerator and the experimental physicists, Rubbia shouted at Rod Cool and the other UA6 physicists that they would not be allowed to make tests. UA1 had been created to do physics, he said, not to make measurements for other experiments. With Ian Butterworth, the CERN director of research, banging on the table and yelling that this was no place to discuss the matter, Rubbia stormed out. He told everyone that if UA6 were to run, UA1 would not.

The next day, the management decreed that they would let UA6 run for the last hour of every shot,* when the beam was at its weakest. It was the time that UA1 had the least to lose by being off. ("That's fine," Rubbia told his physicists, "that way we sleep and they keep working.") After running like that for a few weeks, UA6 could then report back on what physics, if any, they could do under those conditions. Rubbia ordered his physicists to keep the central detector off while UA6 ran, thus ensuring that UA6 could not sneak a test and back their claim that the gas jet was harmless.

Meanwhile, Rod Cool was watching four years of his work, and that of three other collaborators, twelve post-docs, and at least six students, being jettisoned to keep Rubbia's boat from rocking. For ten years, Cool had been director of high-energy physics at Brookhaven. He was a distinguished physicist, who, among other things, had refused the Harvard position that was later offered to Rubbia. He was also renowned as the best poker player in high-energy physics.

Cool called the CERN judgment tantamount to aborting the experiment, reversing a decision that had been made four years earlier. He talked openly about taking it to court, if such a thing were possible. He said, "Mr. Rubbia has made errors before and even the Nobel Prize doesn't mean that he is infallible in questions of a scientific nature, nor that he is God or the Pope." Cool blamed Rubbia for the situation: "Carlo has in his, paranoia, if you like, megalomania, if you like, succeeded in subverting the usual scientific process." And he blamed CERN's Nobel public relations campaign. "It's an unusual circumstance where so much publicity has been given over such a long period to more or less a single individual," he said. "It's questionable whether the Director General is any longer able to say no to him."

Everything seemed to have calmed down until one afternoon

---

*When two beams are in the machine and colliding, it is called a shot. This lasts for roughly twelve hours, until there are no longer enough particles in the beams for frequent collisions.

when Ian Butterworth walked over to Rubbia's office with news of the latest decision regarding UA6. Butterworth was a British physicist, serving a three-year term as research director, the second most powerful position at CERN after that of DG. He was short, a little plump, and he always seemed to look a little harried, like he'd rather be a physicist again instead of a bureaucrat. He had come to tell Rubbia that the SPS Committee—the board of physicists that decides which experiments will run on the accelerator—had voted to let UA6 run for ten days at full power. And that they had done so without asking Rubbia to testify.

Rubbia walked Butterworth out of his office, and as Butterworth backed down the stairs, Rubbia screamed at him that the committee was ignorant, that UA6 could not be allowed to run, that they didn't know what they were doing to UA1 and how this would affect the physics. Rubbia was angry but not yet fully out of control: He still thought he could stop it.

When the tirade was over, Rubbia got on the phone to Schopper. The following week yet another level of the CERN hierarchy would meet to rubber stamp this decision, and Rubbia was telling Schopper that this time he expected to be present.

"They go for a head-on collision with us," Rubbia said later at the six o'clock meeting. "I think they are going to find that my head is pretty hard."

Three weeks later, Rubbia had the decision reversed. UA6 got ten days, but they had to keep their gas jet on at only 10 percent of its full capacity, which would allow them to do little more than debug their equipment. Jack Steinberger put the general opinion of the laboratory succinctly: Speaking of Rod Cool, he said, "Great, you work in physics your whole life, and then when you're an old man, they kick you in the backside."

Once the Nobel hangovers had receded and the UA6 situation had settled, the physicists returned full time to physics. They had already found two new monojets. The first was picture perfect. It was scanned at three-thirty on a weekday afternoon, triple-checked by six that evening, and by ten at night, Rubbia's secretary in Boston had a message for him that they had found the first monojet of 1984. They had also found one Z, but one that nobody particularly liked. On a Saturday evening in late October they found their second Z of the year and the first good one. The event was striking: two beautiful electrons going off back to back.

"Pretty clean, huh?" said Rohlf.

Poking out from the side of the electron pair was a tiny jet. There was nothing else. It took them about one minute to decide it was a Z.

"It's a Z . . . nice," Rohlf repeated. "Nice, huh?"

Rohlf zoomed the image down so that it was no more than an inch across on the screen. When he had it as small as it would go, he began to pivot it with the joystick, spinning it around in space in the middle of the screen.

"Glashow came over to look at the monojet," Rohlf said. "He'd never seen the Megatek before. First thing he says when he saw it, 'You know, this would be a great machine for pornography.' "

As Rohlf told the story, the Z was spinning slowly in electromagnetic space, its electron legs spread for all the world to see, its small jet protruding nakedly.

Two days later, Rohlf showed the two Z's to Alan Norton, a Scottish physicist in his early forties. Rohlf showed Norton the first Z, the one that Rohlf and others hadn't liked, and then moved quickly on to the latest one.

"This one's fantastic," said Rohlf. "It's a good one, huh?"

"Yea," said Norton.

They looked at their Z. Every few minutes, one of them would say, "Isn't that nice?" They savored it like a fine wine. They were looking at something that had been captured only fifteen or twenty times in history. It was worth savoring. To hell with wine. Anyone could make wine.

That night, over hamburgers at Wendy's, Cline and Rohlf discussed the importance of Nobel telegrams. All weekend, Rubbia had been seen pacing the corridors, walking the five feet between his own office and his secretary's, worrying about telegrams. By the time life calmed down, he would have received in the neighborhood of two thousand of them. (The local favorite was from a U.S. physicist who wrote, "Congratulations Carlo, I'm sure this will not affect you and you will remain your quiet and subdued self.") Thank-you notes had to be sent to all. Rubbia was even getting drawings of his experiment from Italian schoolchildren with inscriptions on the back reading: "To the great Italian scientist Carlo Rubbia."

The importance, according to Cline and Rohlf, lay as much in who didn't send telegrams as who did. To have worked with Rubbia in the past and not send a telegram of congratulations was to say that you did not care if he was on his way to becoming about the most powerful man in physics, as well as the future director general of CERN. And if you wanted to make up with Rubbia for something that happened in the past, the way to do it was with a telegram, because Rubbia was likely to forgive, at least in public, past animosity.

"Carlo looks at those things," said Rohlf.

"He wants to know," said Cline, "if the people he hated six years ago, he should still hate."

Cline was worried. He had just returned from the States, and he had found that the two-jet event with the missing energy, the dijet, was no secret in America. The supersymmetrists were already climbing all over it. As far as Cline could figure, one of the Americans in UA1 had told a Wisconsin theorist about the event, and it had spread outward from there. Rohlf added that one of the German physicists had also flashed the event at a lecture at DESY.

"Oh, no," said Cline. "It's not what you say, it's what you appear to say. Now we're in trouble."

Unfortunately, Cline no longer believed that the event was real. They had good reason to think that the machine had faked the event, that the huge missing energy had been computed because some particle had escaped into the cracks in the detector. Even Rohlf was skeptical now.

"It's all over America," Cline said glumly.

Late the following night Rubbia sat in the Megatek room chatting with Cline and Rohlf. Rubbia planned to leave the next morning for Santa Fe and an annual American Physical Society meeting that would be attended by well known physicists such as Gell-Mann, Lederman, Ting, and Richter. Rubbia had asked Rohlf and Cline to gather all their new events of the year, the new W's, Z's, and monojets.

Rubbia leaned back in his chair, relaxing. Rohlf leafed through sheets of printouts looking at the events, while Cline collated. Rubbia was in an expansive mood; he smoked a cigarette, folded his hands behind his head, and aired what he was thinking of saying at Santa Fe. "We've got these monojets," he told me, "seven of them, and we can't explain them. There's no way to explain them. It's got to be a new particle. We're stepping from the known to the unknown. We can't lose." He then digressed to tell the story of one of his male physicists who danced mistakenly with a man all night at a Genevoise strip joint, but Rubbia broke off in mid-story to get back to work. On his way out, he stopped at the door.

"Have you heard about this memo?" he asked me.

No.

It was a memo from the director general, "Sir Alec Morrison, président du conseil du CERN, et le professeur H. Schopper, directeur-général," inviting CERN physicists to a reception for "Messieurs Carlo Rubbia et Simon van der Meer." Rubbia found it antagonizing, but not sufficient to ruin his good mood. "This is the way these people think," he told me. "Sir Alec Morrison, le professeur H. Schopper, and then Mr. Carlo Rubbia. Mr. . . . Put that in your book. It's perfect."

He smiled. The next day Rubbia left for New Mexico and the memo was tacked to the bulletin board outside his door, the offending portions highlighted.

11.

NOVEMBER, 1984: THE TAIL OF THE DISTRIBUTION

"Even the heads of laboratories don't like to put their heads above the parapet."   —a British physicist from UA1

It's early November, a typical night in the control room of UA1. Cline, recently back from the States, is working a midnight to eight shift on the muon chambers. His brown hair is thick and hanging down below his collar in back; his sideburns are equally long and bushy but going gray, and he is wearing black-framed glasses. He is drafting a letter to NASA about his latest idea: putting a gravity-wave detector on the next space station, should one ever be built. Every few minutes he glances at the dozen or so computer screens in front of him to make sure that the chambers are running, and then goes back to his letter. At three in the morning Rohlf comes over from the scanning room. He's found another couple of W's and a beautiful monojet, and now he wants to hear the news from the Sante Fe meeting. They stand at the coffee machine out in the corridor, drinking espresso from plastic cups stirred with little plastic straws, while Cline fills him in.

First item: The Zeta was officially reported dead. No more rumors. The funereal talk was given by a young post-doc from the experiment.

Second item: Rubbia was mad at Sam Ting. Ting had lectured about his new detector for the Large Electron Positron machine, LEP, CERN's next huge accelerator venture, and had suggested that the second stage of LEP would be another collider in the same tunnel, called the Large Hadron Collider (LHC). This would be a scaled-up version of Rubbia's present collider, twenty times more powerful. Ting had claimed that his detector would work fine for that, too. And he had shown a telegram from Schopper that either Ting or Rubbia had construed as giving Ting official permission to talk about it. But the LHC was Rubbia's baby. Since the first collider was his idea, he considered that he had proprietary rights to the second, too.

When Rubbia got back to CERN, he demanded that Schopper explain this telegram. It turned out to be innocuous. All it said was that the LEP start-up was on schedule for 1988. Somehow Rubbia had misconstrued it. "A complete misunderstanding" was what Schopper called it.

"Carlo doesn't know what to do now," said Cline. "It's hard to talk with enthusiasm about the Large Hadron Collider when Sam is sitting in your way."

Third item: Rubbia gave the after-dinner speech at the conference banquet, and it didn't go over well with the American physicists. Leon Lederman had introduced him with an old joke that he twisted for the occasion. Although Lederman thought it was a "friendly joke," Rubbia apparently did not.

Rubbia made two points in his after-dinner speech. He talked about the Americans' plans for the Super Conducting Supercollider and, as Cline related it, "Carlo said that building the SSC is harder than going to the moon. Which the Americans didn't like. And second, there should be a collaboration with Europe, intellectually, if not otherwise. Which they didn't like. So his after-dinner speech didn't go very well."

Item four: "The only interesting thing at Sante Fe was UA1."

Throughout November, the collider ran like never before. It was as if they had hit the mother lode of physics. The scanners would come up with two Z events in an evening, and the SPS control would flash a message on the television screen saying that they were going to keep the beam until after the integrated luminosity per shot reached 10 inverse nanobarns. They had had only 28 in the whole 1982 run, and now they were nonchalantly saying they'd drop the beam after they hit 10 in a day.

The next day was the same. The physicists spent the time watching the inverse luminosity count climb up on the screen. By about seven in the evening it was hitting 10 inverse nanobarns and by the next morning it had tapped out at 17. By then they already had a record stack of antiprotons stored in the accumulator ready for the next shot, nearly twice as many as they had had two nights earlier. In the UA1 six o'clock meeting the physicists joked that this should go in the *Guinness Book of World Records:* the most antiprotons ever put in one place in this part of the galaxy, and maybe in the universe.

Every time I checked in at the Megatek room, either Rohlf or Felicitas Pauss or one of the other physicists would look up and say, look at this, and then show me some bizarre event that they'd never seen before. And everyone would smile. The machine was running so well that the collaboration was floating on air.

The only catch to this wonderful performance was that the detector wasn't built to handle it. The physicists were getting the best beam they'd ever had, but the collisions were coming so rapidly that it was as if the physics was being rammed down their throats. The six o'clock meetings often deteriorated into sometimes vicious, sometimes amusing arguments about how to handle all the collisions. And more often than not the arguments seemed to pit the majority of the collaboration against the small group of Germans who had built the muon chambers.

The Germans were led by Karsten Eggert, forty-one, with a pronounced chin and gray-blue eyes that were lost behind lids that could have used more sleep in the past decade. He smoked a pipe, and like most everyone else at UA1, talked as though he never had enough time to say what he wanted. Eggert had grown up in northern Germany. His father had died in the war, and he had turned to physics despite family pressure to go into industry. "Somehow I had the feeling," he told me, "that if you go into industry, you have to do a lot of administration and a lot of business, whereas pure research was not really a business." Since then he had learned differently.

Eggert had been fighting for his muon chambers since they'd been installed three years earlier. The arguments started when the massive muon chambers—4,000 square meters covering the sides of the detector, the top, and the front—got blasted by particles coming down the beam pipe and threw the electronic equivalent of conniption fits. When a muon shot through the chambers, it left an electronic signal which would then be boosted by electronic amplifiers. But the extraneous particles that spewed out of the collision down near the beam pipes occasionally fired an amplifier, which would set off other nearby amplifiers, which in turn would ignite others, until it started an avalanche. In no time at all the detector would be triggered again and again, because the muon chambers kept reporting that they were seeing something interesting when in fact they weren't.

While the chambers were twitching, the data being taken were junk. If anything interesting happened, the detector would be too busy storing the junk on tape to notice.

The physicists fought daily over these oscillations. Even when the muon people, as they were called, were in the right, and the oscillations were not their fault, nobody believed them. In the six o'clock meetings, Eggert or one of his crew would suggest that some other piece of the detector was culpable—once they claimed that one of the computers was triggering the avalanche—and someone else would say, "Sure, Karsten, maybe it's the coffee machine in the hall that's doing it." And the collaboration would collapse in hysterical laughter.

On one occasion a UA1 technician rolled his car into the building and hooked the muon chambers up to his CB radio, blasting them with static to see if they would oscillate. Then the pompiers—the CERN firemen—tried to rock the chambers with the transmissions from their walkie-talkies. But the chambers didn't oscillate. It turned out the next day that it really was not the muon chambers but one of the photo-multiplier tubes in the calorimeters that was the cause of all the problems. The muon people were innocent.

But by then they had more serious problems. The detector, no matter how well it was running, could record the details from only three collisions every second onto the computer tape. That was three out of eight thousand collisions per second. The better the accelerator ran, the higher the luminosity of the beams, the more collisions they would have, and the more they would want to save.

Triggering on muons had the advantage that even when the muons were of relatively low energy, the detector would have no trouble recognizing them when they passed through. Electrons, on the other hand, became distinctive only at higher energies. The physicists could do good physics with both muons and electrons, but they could trigger on only one at a time. The muon people wanted to look for W's and Z's and new physics with muons. But the rest of the collaboration preferred electrons.

The collaboration had spent the past year developing complicated software and hardware systems that were more and more selective about what they considered a good event—or more knowledgeable about what they knew to be garbage. But at the same time, the accelerator was running better than ever before, and it kept overloading the system.* And again it seemed to be the muons that were the culprits. When the system did overload, the

---

*This was a worry for the future, because the next generation of accelerators, the SSC and the LHC, would create maybe one hundred times more collisions each second.

physicists would simply remove the muon chambers from the trigger, so that the chambers would still record physics but the muon chambers couldn't trigger the experiment.

When the beam came in and the trigger rates went wild, the physicists in the control room would begin to panic. Someone would yell to cut the muons from the trigger system. Then Eggert would fight back until finally another physicist would call Rubbia in his office or at home. Rubbia would yell, "Yank 'em," and the muons would go. But until then Eggert would stonewall, knowing that as long as he could keep them talking, or listening, and not yanking, he would have data coming in.

Giorgi Salvini, an elder statesman Italian physicist, was in charge of the control room one night when the trigger rate shot up. He had them pull the muons immediately. He said later, "It was like ripping a baby out of its mother's arms."

By November 15, just as the UA1 physicists were bringing the trigger rates under control and life was beginning to calm down for Eggert and his colleagues, Antonino Zichichi showed up. Zichichi was an Italian physicist of about Rubbia's age. He was small and perfectly groomed, with thick gray hair that he brushed back and grew almost to his shoulders. He had a charming smile and a very dapper manner, and tended to wear beautifully cut Italian suits. Zichichi was Sicilian—his father was a businessman in Palermo—and the most powerful figure in Italian science. He wrote a weekly newspaper column, was constantly lecturing and appearing on television, and he had almost singlehandedly raised the awareness and interest among Italians in the nuts and bolts of physics beyond that of any other nationality. Zichichi had also built the largest underground laboratory in the world, called Gran Sasso, beneath a mountain sixty miles east of Rome, and he was much beloved by the locals, whose economy he had stimulated. Cline tells the story about how Zichichi was walking down an Italian beach one day with Glashow, and a photographer came up and asked Glashow, the Nobel Laureate, to step out of the way, so he could get a picture of Zichichi.

During the sixties, Zichichi and Rubbia were rivals at CERN. They were the two young hot Italian physicists, and they both had a very competitive, very ambitious style of physics. Victor Weisskopf, who was director general of CERN during the first half of the sixties, called Zichichi and Rubbia condottieri types, after the ruthless mercenaries of the Renaissance. "One day," Weisskopf related, "comes these two characters, Zichichi and Rubbia, into my office, screaming at each other. Zichichi said, 'Carlo has put his car in my parking place.' Carlo said, 'Nino has put his car across my car. It's my parking place.' I said, 'There is the door. You get out immediately. Coming in with such things!'"

On another occasion, Weisskopf received a call from an Italian journalist. "He asked me how it was to be a director general of CERN and was it difficult to have different nations working together? How can you make Britons work with French and Germans? And I said, 'That's no problem whatsoever. The real problem is to have two Italians work together.'" Weisskopf liked to say that there was so much screaming in Italian at the lab that he would get home and ask his wife, "Am I director of CERN or of La Scala?"

Still, Zichichi had done good physics in the sixties ("In 1965," he told me, "I discovered the Deuteron"). He had missed the J/psi at Frascati in the early seventies, as well as the tau lepton ("If the tau lepton had existed in the 1 GeV range, I would have discovered it"). Then he had moved to the ISR. Now he wanted to get into UA1, to do muon physics.

Zichichi wanted to take Rubbia's UA1 detector and install new apparatus to detect muons. UA1 already had Eggert's chambers to catch muons. But they did not cover the area within 15 degrees of the beam pipe, where the detector was overwhelmed by the particles coming down the beam pipe and the remnants of the incoming protons and antiprotons spewed out from the collision. Zichichi wanted to put his chambers down in that swamp, and he swore he could cull physics from it. His equipment would ignore the junk and signal only when a nice muon had shot from the collision. Then he would be able to trigger the entire UA1 detector and gather all the data collected there. Zichichi wanted to sit piggy-back on UA1. He had originally suggested this idea to UA2, and they had more or less said that they would not have any room for his chambers.

Rubbia was informed of Zichichi's plan in an SPS Committee meeting. The committee was backing it. "The committee was very happy," Rubbia said later. "What a pleasure to throw a monkey wrench into our system after UA6." On November 15, Rubbia broached the topic to his own executive committee; "We have a danger of having a Sicilian-Libyan component in our group," he told them.

Rubbia first suggested that they stall Zichichi by agreeing to his requests but claiming that the cost of what he hoped to do was enormous. "Make it so expensive that people have to think twice," he said.

"Is their physics good?" asked John Dowell, head of the Birmingham group.

"Yeah," responded Rubbia, "but, of course, physics is irrelevant to that committee."

Eggert sat in on the next executive committee meeting and listened to Rubbia say that Zichichi wanted to step in and do muon physics at UA1. To Eggert, Rubbia did not seem to be taking as strong a stand as he might. Eggert had thought for many years about putting chambers where Zichichi proposed, but he had not yet come up with an acceptable design. He suggested they tell the relevant CERN committees that they were already planning to do exactly what Zichichi had suggested, which in a sense they were.

"You can't do this," Rubbia replied. "You can't say you want to build Zichichi's proposal yourself."

"It's easy to show that Zichichi's proposal will not work on us," Eggert said.

"Karsten, are you volunteering to do the proposal yourself?"

"Yes."

"They will not take away an experiment proposed by Zichichi and give it to you," Rubbia said. "The committee has endorsed his document entirely. They have the right to trigger your detector and take data from you, and you will have the right to work for him, and if you screw up, he can kick your ass."

"Why don't you say it's impossible to go lower than fifteen degrees," suggested one physicist.

"But it is possible," Eggert insisted. It was Zichichi's particular proposal, he thought, that would never work.

"But if you say it's possible," Rubbia said, sounding like a patient mother explaining the facts of life to her errant boy, "Zichichi will take your proposal and say, 'Thank you very much, you have done our feasibility study.' What you can say is, 'Yes, it's possible, but it will cost twenty million Swiss francs.'

"I will never sign a paper with this guy," Rubbia concluded loudly. "Let's have him here . . . I would love to see Karsten and Zichichi arguing about muons. That would be funny."

Rubbia's stance on Zichichi was uncharacteristically ambivalent. He was simply not fighting with his usual passion to keep Zichichi off his back. But his group truly seemed to dislike the idea of working with Zichichi. He had a reputation in the sixties as one of the three toughest men to work for at CERN. (Rubbia was another. The third was also Italian.) And some of the group had worked with him before, and had not liked it.

"If the majority of UA1 does not want to work for Zichichi," Rubbia told me, "I'm sorry. That's a problem between Zichichi and them. Really, this is not my responsibility." Several of the UA1 physicists suggested that Rubbia was being politic by taking such a line on Zichichi, who was too powerful in Italy to cross. Even though Rubbia, with his Nobel, was perhaps more influential in political circles, Rubbia needed him as an ally. Cline, on the other hand, had told me a month earlier that Rubbia and Zichichi were making a deal: Rubbia would let Zichichi into UA1 if Zichichi would back Rubbia's plans to put a huge proton-decay experiment into his underground laboratory at Gran Sasso.

The night after the executive committee met, Cline claimed that Zichichi would be a member of UA1 "only over a lot of people's dead bodies." And when I asked if that meant that Rubbia would not put his experiment in Gran Sasso, Cline said, "We're going to sell it to Sudbury, this 7,000-foot-deep mine in Canada. The Canadians want us badly. Gran Sasso can shove it up their ass." But this did not exactly seem to be the case.

At the center of the controversy seemed to be the proton-decay experiment, which had a long and twisted history, and which now seemed to be insinuating itself into the politics of UA1. The concept of proton decay had surfaced seriously in the mid-seventies. Glashow, Salam, and Weinberg, among others, had been working on their grand unified theories (GUTs), which attempted to marry the strong force and the electroweak force. The equations of these GUTs implied that protons need not live forever and therefore—by the laws of probability—would decay, being snuffed out in infinitesimal bursts of energy after an average lifetime of a million trillion trillion years or so. This was not only a racy idea, but, bizarre as it may seem, the sole prediction of the theories that seemed to be within the realmofexperimentalconfirmation.

At about the time Rubbia was getting the collider rolling at CERN, Glashow, Weinberg and Nanopoulos, a post-doc at the time were talking to Larry Sulak, assistant professor at Harvard, about an experiment that would search for proton decay. Sulak had worked for Rubbia at Fermilab and had later helped set up and then run Rubbia's Brookhaven experiment. Sulak and the theorists considered methods by which they could enclose a million trillion trillion protons in one place and then watch them closely enough to catch one decaying. Eventually, Sulak hit upon an idea that had been floating around the field for quite a while: simply use a huge tank of water that would contain the necessary protons in the form of the hydrogen atoms in water molecules. Such a tank would be placed in a mine as far underground as possible to shield it from cosmic rays, and surrounded by sensitive light meters that would catch the infinitesimal flashes of light marking the death of a proton.

The relationship between Sulak and Rubbia had been rough at the time. Rubbia had asked Sulak to help him with the collider at CERN and, for personal reasons, Sulak had refused. When Sulak went to Rubbia with the idea of working together on a water detector for proton decay, Rubbia claimed he'd been thinking of the same thing. ("Carlo insists to this day," Cline told me, "that Sulak had nothing to do with it; the idea all was his.") And instead of working with Sulak, Rubbia chose to compete with him.

At this time Sulak was refused tenure at Harvard and turned to the University of Michigan for both tenure and support on the proton-decay experiment. Rubbia wrote a letter of "dis-recommendation" for Sulak that has become renowned among Harvard physics alums. Sulak was shown the letter by one of the Michigan physicists. As he described it, "Essentially everything that Carlo had done wrong in the previous eight years was attributed to me. Many of these things I didn't have anything to do with. It was explicit as to how I had screwed this up or screwed that up." Sulak received tenure in spite of the letter, however, being helped considerably by recommendations from Glashow and Weinberg, and also by a telegram from Glashow suggesting that Michigan ignore one of the letters from Harvard since one of his colleagues "might be mad."

In December, 1978, Dave Cline organized a proton-decay workshop at the University of Wisconsin. After the interested parties had presented their papers, Cline published a volume of conference proceedings that included every talk except Sulak's. Sulak told me he would never forgive Cline for this. However, Cline later volunteered that it was Rubbia who had insisted that he withhold Sulak's paper. When I asked why, Cline said, "Carlo has a very good smell for competition, and he probably felt that Sulak would be our competition."

Sulak eventually hooked up with Michigan. and together with physicists from Brookhaven and the University of California at Irvine (their collaboration was known as IMB) submitted a proposal to the Department of Energy for a water detector to be placed in a salt mine near Cleveland. Rubbia and Cline submitted a proposal for an even larger experiment for a Minnesota mine that, as Cline told me, "We could have built in principle. But we were misled about the mine. It was collapsing."

The DOE sided with the IMB group.

Cline and Rubbia then turned around and proposed, with Jim Gaidos of Purdue (their collaboration was known as HPW), a quick and dirty experiment for a Utah mine that would be smaller than IMB's but that could be built quicker. This one the DOE accepted. "Suppose the proton-decay lifetime was indeed short," Bernie Hildebrand of the DOE told me, "and HPW found it first. Okay, another Nobel Prize for Carlo and down the drain for IMB. But then IMB would still have all kinds of decay modes which had to be studied."

Both experiments were built and taking data by 1982. By January of 1984 it was clear that the proton-decay lifetime was not within range of the experiments. Sulak's IMB experiment set the definitive limits on the nonexistence of proton decay and, in so doing, disproved the simplest and most likely grand unified theory. (This defeat for the GUTs, among other things, set the theorists to looking seriously at supersymmetry.) Cline and Rubbia's HPW experiment, which the two of them rarely if ever visited (according to Hildebrand, it would "have been all the way down the drain" if not for Gaidos), was, as Glashow put it, "an unmitigated failure."

Now, in November, 1984, Rubbia was again looking into proton decay. He thought the time had come, even though most physicists believed that if proton decay had not been found already, it was unlikely that it would ever be found. At Harvard, Rubbia had his post-docs and grad students working on liquid argon for use in a proton-decay detector. It would, so he said, allow them to build a huge detector, much more effective than the IMB detector, which would give them pictures of proton-decay events as good as those from the central detector of UA1. "We are the only ones who know how to build a liquid argon detector," he claimed, "in the same way we were the only ones in the late seventies who knew how to kill a Z. I have a generation of students sweating liquid argon tears, and now it works."

That Rubbia was off on an underground expedition was indisputable. Now the only question was where he was going to put it. He had Cline look into the mine in Canada, which Cline said was perfect. Sulak told me it was "a terrible place. It's spooky as hell. Everything's sweltering." Rubbia had met with Zichichi off and on about Gran Sasso, and Cline was even a member of the scheduling committee of Gran Sasso. But nothing was settled. Rubbia seemed to be playing his cards as safely as he could. Stalling, as he put it. Karsten Eggert and the UA1 physicists could only wait to see which way the decision went.*

Dave Cline, while doing the legwork for the proton-decay experiment, was still filling his occasional late night shifts at UA1. But Rohlf told me one afternoon that Cline had lost his confidence in the monojets, and believed that they all looked like background—boring old standard model physics. He had said to Rohlf, "Don't tell anybody because it might cause a panic." So I went in search of Cline and found him a few days later doing his midnight to eight muon shift in the control room.

I told him there was a rumor that he had become a disbeliever. "Who says I don't believe in monojets?" he immediately asked. Then he answered his own question. "Probably Jim Rohlf. He's being too emotional about this. It's not that I don't believe in it. It's that I don't think we've carefully calculated all conceivable backgrounds for this. And I think it's time to go back and do that."

Cline's fears were strung on a tight wire between his past and the prospects for the future. A month earlier, I had spoken with him about the neutral currents and the high-y anomaly; he had admitted that he hates even to hear the word "anomaly" anymore. He said they got in trouble on the high-y by talking about it too much.

"We gave people the impression that we knew this was new physics," he said, "very much in the same way that we say the monojets is new physics."

More than anything it was Rubbia's attitude toward the monojets that had Cline anxious. "In fact," he said, "Carlo told me the other day, 'Look at this, we're gifted. In 1973, we discovered neutral currents. In 1974, we discovered dimuons. In 1983, we discovered the W and the Z. In 1984, we discovered supersymmetry.' " This

*By October, 1985, Rubbia's $20 million proposal had been accepted at Gran Sasso, and had been promised funding by the Italian government. He was competing with several other proposals, including one involving Sulak, for U.S. dollars. Zichichi had set up small detectors in UA1 to begin the first phase of his experiment.

made Cline think that Rubbia had made up his mind already.

When Cline speculated about the future of the field, he found more to worry about. He reminded me of the old line that hadron colliders—proton-proton or proton-antiproton—would be like two garbage cans hitting each other, and how Rubbia had to some extent disproved this with the discovery of the W and the Z. The clincher had been the new physics, which included the monojets and the tentative claim that they had captured the top quark. Rubbia had convinced the physics community that hadron colliders were clean enough to do anything. Now the machines of the future, the semi-billion-dollar LHC and the $5 billion SSC, would be hadron colliders. "They just altered people's perceptions about the future of the field," Cline told me. "But I think it's a very dangerous situation we're in now. We can alter the course of the history or we can make a false start and get people going off in the wrong direction."

If the monojets were not a breakthrough—as Rubbia had been telling the world they were—it meant they were simply standard model physics. When the monojets had appeared the previous winter, everyone had conceded that they could be faked by a Z decaying into an antineutrino and a neutrino. In such an event, just before a quark and an antiquark hit, the quark, for instance, would split off a gluon, which the theory of the strong force allowed for naturally. Then the quark would recoil in the opposite direction and collide with the antiquark. This would create a Z, which would decay into two neutrinos. Since the neutrinos are invisible to the detector, that would appear to the detector as missing energy. On the other side, the gluon would fragment and create a jet of particles. The physicists would see the jet from the gluon and nothing else but missing energy—a monojet.

But they had calculated the frequency that this would happen, using the rough guidelines of quantum chronodynamics, QCD, and the result seemed to be not nearly sufficient to explain the six monojets they had found. They had a much trickier suspect, however, which was called W to tau decay. Here the quark and antiquark created a W, which decayed into the heaviest of the electron family, the tau, and a neutrino. The neutrino disappeared; the tau then decayed, according to theory, into either one particle or three particles, and showed up like a jet in the detector. If this happened, what they would see would be the missing energy and the jet. The tau was also a type of monojet, and when Rubbia referred to monojets it was a little misleading, for what he really meant when he said monojet was a monojet that had too much energy to come from a W to tau decay.

Theoretically, they knew that if the jet had an odd number of tracks, it was probably from a tau. If the missing energy was a little less than half the W mass, then that also meant they were likely seeing a W decay.

If only the universe was that simple. If one of the original quarks or antiquarks had ejected a gluon just before the collision, then the W, when it was created, would be given a boost away from the departing gluon. If the W decayed and the neutrino shot out in the direction of the boost, that could increase the missing energy making it greater than half the W mass. And then the tau jet would go the way of the gluon jet, the two might get entangled, and the physicists might see any number of tracks in the jet—odd or even.

This was a complicated mess; the only way they could deal with it was to crank meticulously through the calculations and pray that the equations of QCD—which provided the probabilities of having a gluon emitted of a particular energy—were accurate. They had done these calculations for the paper the previous year, but the calculations had been rushed, because Rubbia had rushed the paper.

And Cline was no longer convinced.

In the previous year's monojet paper, they had labeled the monojets with letters from A to F. A was the most spectacular, the one that Mohammad Mohammadi had found. F was the most mundane, and they admitted that it was likely to have been from a W to tau decay. As far as Cline was concerned, only events A and D were definitely not W to tau decays. He had yet to see anything this year about which he could make the same claim. "The fact is," he explained, "the new events in some ways look even more like taus. That's the disturbing thing."

Cline did not consider himself an innocent bystander in all this, and he was nervous. He had had Mohammadi, his graduate student, work on the background calculations the previous year. "I pushed him into this," Cline said, "so I feel particularly nervous that if this result is wrong, he will somehow be blamed."

Mohammadi was twenty-nine years old, tall, slender, and handsome. He had short thick black hair, and usually a couple of days' growth of beard. Like everyone else at the lab, he looked like he could use a couple of months' sleep. His family was Persian and he had left Iran back in 1973 to get the best education possible, which meant going to the States. He spent a year in Missouri, because he had friends there, but found the culture shock too much to deal with, and moved up to Wisconsin. He hooked up with Cline for his Ph.D. and worked with him on pbarp at Fermilab before moving first to Berkeley, then to CERN in February 1983.

Mohammadi had started his work on speculative electron-like particles called heavy leptons, which would be heavier even than taus. Cline had always seemed enamored of the concept of heavy leptons—the high-y and the trimuons were both misinterpreted as possible heavy leptons—so he set Mohammadi to work on them. If a heavy lepton existed, conceivably the W would decay into it. When the monojets appeared a year later, Jean-Pierre Revol asked Mohammadi if he'd like to pitch in with the analysis, and he said yes. The two were already friends—they were two of the small number of UA1 physicists who believed that entertainment outside the laboratory should be pursued with as much zest as work inside.

Virtually any night you could find Mohammadi in the computer room, slumped in front of a terminal, nearly buried under stacks of computer printouts. On the door behind him there was a poster of Whistler's Mother sitting at a computer, with the caption: "Home Is Where the Computer Is."

One night in late November, I found Mohammadi redoing the tau background calculation. He was, as he put it, beating it to death, feeding into it what he considered even the most trivial and mundane and inconsequential effects, none of which he had been given time to put in the previous year. He was working on what was called a Monte Carlo program.

The Monte Carlo was one way to assess the background to a new signal, a new particle you thought you might have seen. It was a computer program into which you fed in everything you knew about the universe—or that portion of it with which you were concerned. You had equations from the theories that determined how particles interacted, how likely they were to be formed, at what angles they were most likely and least likely to come shooting out of a decaying particle, in what directions they would most prefer to spin, how massive they would be, and how much energy they were likely to have. You also had all the experimental evidence from the past several decades, which gave exact numbers for many of the interactions.

When you came to the part in your program that described the characteristics of a particular particle, you might feed in numbers that had been taken from an experiment run in 1967 at an atom smasher long since closed down—if those numbers were still accurate. Or you might use numbers that had come out of UA1 or UA2 the previous year. The Monte Carlo would contain all the information about all the particles, with all the different probabilities as defined by the theories and the experiments.

When you set it in motion, the computer would take an imaginary proton and collide it with an imaginary antiproton at a particular energy, and would simulate the collision. It would make imaginary new particles that would decay into more imaginary particles, and these in turn would send imaginary debris hurtling into an imaginary detector. Each time the computer had to choose a mass or a direction that was determined by a range of possible numbers, it would essentially roll dice, choosing one of the numbers at random—hence the name Monte Carlo. All of this would be done so fast that you could create a billion imaginary collisions in the same way the collider created a billion real ones.

The more completely you fed the accumulated knowledge of nature into your Monte Carlo, the more accurately the program would imitate nature. When it was done, if the collider was giving you effects that just did not show up in the Monte Carlo, then either you had screwed up the Monte Carlo, or maybe, just maybe, you had discovered something new.

Mohammadi had been working on his Monte Carlo on and off for a year. Now he was adding to it all those inconsequential effects that in the lingo of physics are known as the tails of the distributions. These are the phenomena that could happen, according to theory, but are so improbable that the chances of their happening are almost zero, yet never quite exactly zero.

The tail of a distribution represents something that statistically is very rare. If you take a bell-shaped curve, for instance, the curve of population versus IQ, at the highest part of the curve is IQ 100, the average. Off to the right the curve slopes down and then levels off at almost but not quite zero. Maybe a few hundred thousand people have IQ's of 140, a few thousand of 170, a few hundred of 190, and a handful above 200. These few represent the tail of the distribution. Even though you might live your life without ever meeting one of these intellects, they exist. The line never really gets down to zero, it just fades slowly away in a long, thin tail.

The monojets and the W's and Z's existed on the tail of the distribution of the standard model. They were created in a few of every billion collisions between protons and antiprotons. With so few events, the danger was that the physicists might misinterpret what really was the tail of the distribution of the standard model as being something entirely new beyond it. (If you searched among the four billion people in the world, you might find a handful with IQ's over 250. They would be truly peculiar representatives of humanity, however, and unless you thoroughly studied the rest of humanity, you would have no idea whether they really fit into the theory of evolution, or whether you should take them as evidence of something else.) The UA1 physicists had two choices: they could log enough collisions so that instead of a handful of events, they had fifty or one hundred—and that would take several more years of running—or they could create a Monte Carlo that would simulate the necessary number of collisions.

At half past midnight on this particular Saturday morning, Mohammadi was playing with a subprogram that would take any particle and decay it according to the rules of the standard model. The program itself, when printed out in very small print on paper, was about an inch thick. "Pages and pages of code," said Mohammadi. His particular program took those particles that came directly from the W to tau decay and then superimposed them on the other junk that would be created from the collision—the other remnants of the proton and antiproton. Mohammadi had run off one sample decay that listed 191 different particles. Seventeen were the remnants of the W decay, and the other 174 were junk from the rest of the collision. He was checking for bugs in his program before he set the Monte Carlo to running a large number of events.

Mohammadi hoped to create a program that would give reliable probabilities for a W to tau decay producing missing energy greater than half the W mass in the UA1 detector. It could only happen with that extra gluon jet, and QCD gave probabilities for how often a gluon with a certain energy could be expected to appear. But the higher the energy in the jet, the less likely it was to occur. A jet energetic enough to give the boost to the W and the tau so that the resultant missing energy would be large enough to be construed as caused by new physics—even though it wasn't—would occur only once for every one hundred or so W's.

But the theory was not precise. Nobody knew for sure what the tail of that distribution looked like. Mohammadi was trying to compensate for that uncertainty by sticking even the most inconsequential-seeming events—all the tails of all the different distributions that related to this particular corner of the universe—into his Monte Carlo. Together, they might add up to something consequential. The year before, when Mohammadi had had three weeks to write his Monte Carlo, he had been told to cut all the tails off, and they had hoped it wouldn't make any difference.

Mohammadi programmed his Monte Carlo to churn out five hundred decays of the W. He typed "Run" on his terminal and then sat back to wait. Meanwhile, he explained the one fatal flaw in a Monte Carlo: that it was only the sum total of everything the physicist knew, but no more.

"If you see something in your data which you cannot reproduce in your Monte Carlo," he said, "it indicates that there is something that you do not understand. It could be something new and it could be something very old. You can only trust the Monte Carlo up to a certain point. But you can check it, compare it with data. For instance, we can generate W's with the program and look at various plots that we have from real data. For the W, it really matches the thing very nicely. The W's are W's; the Z's are Z's, which means we really know the standard theory. Now, in the case of monojets, we don't have a Monte Carlo to generate monojets, so we're stuck in a way."

The program finished. It was two-thirty in the morning. Mohammadi decided to look at two last numbers to make sure that, as he put it, "we haven't generated junk." He had written them into the program as a triple check; the first should always be larger than the second. It wasn't.

"What happened?" he muttered to himself.

I decided to go home. I had reached the point of diminishing returns. As I walked out the door, Mohammadi said to me, "This could be done in five minutes, or by tomorrow afternoon."

The life of a graduate student.

Physicists have a natural aversion to basing their case for some new piece of physics on a Monte Carlo alone. The Monte Carlo is limited by what they know about nature, and that limit is reduced even more by whatever shortcuts they take in transferring that knowledge from brain to computer. This means that they can never prove conclusively, either to themselves or anyone else, that they haven't concocted a signal by leaving something out of the Monte Carlo.

Much of the creativity in the analysis lies in improvising tech-

niques for gauging the background and the signal without needing
to fall back on your Monte Carlo. For the monojets, this meant
making the intuitive jump from those events recorded in the machine that they recognized as being from the tail of the distribution
—like the obvious W decays—to those background phenomena that
nature must be throwing at them from the tail, but which they would
not recognize—like the much less obvious W decays. Normally, this
would be the preferred method of computing background. But being
out on the tail, they never got enough of any phenomena to make
sufficiently strong speculations about how much of something they
had seen, and how much of something else they should expect to
see.

The previous year they had had a handful of bizarre events. If
they had been lucky, it could be new physics. That was what Rubbia
had put his money on. If they had been unlucky, it was the tail of
the distribution raising itself farther out of the swamp than expected. They would need more data to bulk up their statistical
arguments. But while these data were on their way, they could begin
to play with what had already arrived.

This was how Daniel Denegri kept himself occupied during the
run. Denegri was Yugoslavian by birth, although educated in Paris
and the States, and he now worked at Saclay. He wore gold-rimmed
glasses, had short black hair and a neat black beard and mustache.
He was almost as energetic as Rubbia, and when he talked about
physics, again it came out like water from a faucet that couldn't be
turned off. One day I interviewed Denegri over lunch, and he
wouldn't eat, because he didn't want to slow down his talk of
physics long enough to chew his food.

Denegri would take the high-speed train, the TGV, down from
Paris, work all hours, day and night, for three or four days, write a
few memos on physics that he would duplicate and pass out to the
entire collaboration, then train back to Paris. Some of the collaboration ignored his memos, claiming that he wrote them faster than they
could read them. Others read them because, as Rohlf said, "he
usually has something to say."

Ever since the fiasco with the jets way back in 1982, Denegri had
become something of a devil's advocate at UA1. Rubbia still blamed
Denegri for the jets, and saw him, or so Rohlf said and Denegri
believed, as more of a nuisance than anything else. Denegri would
come up with some new argument or twist on how Rubbia's new
physics breakthrough could be boring everyday background, and

then try to present it in a meeting, or walk into Rubbia's office with it. Rubbia more often than not would explode. And then Denegri would disappear for another few months. Every experiment needed a devil's advocate, if not many, but Denegri seemed to bug Rubbia so much that it was doubtful whether Rubbia ever listened to him.

Denegri's latest memos had claimed that of the handful of monojets in the monojet paper, he could write off all but one as compatible with background. As one Danish physicist put it, "He's really hilarious. He starts with five events but he still says, don't worry. There's still one that you can't explain away." Rubbia never found it hilarious.

Some of the original work on monojets had been done at Saclay, but they had never felt certain that they were looking at anything but W to tau decays. Michel Spiro, the head of the group, had been most vociferous on this point, and indeed had quit over it. Denegri had considered removing his name from the paper, as had Spiro, but had decided against it. He was also working on the analysis for the top quark, and he felt that if he blatantly disagreed with Rubbia, he would destroy what was left of any working relationship between them. He signed.

Now he seemed to have dedicated his life to proving that the monojets were junk.

At eleven on a Wednesday night, I found him in the copying room of the office block, madly duplicating yet another memo. I asked him if he had anything interesting, realizing immediately that this was a remarkably stupid question. Denegri took off at his usual hyperactive pace, telling me that he had another explanation for the monojets. From a pile of papers, he whipped out a Megatek picture of a W they had spotted just that day. It had an enormous 40 GeV jet with it, from an ejected gluon. They had never found jets that large with W's before, and had thought them almost impossible.

Then Denegri explained to me how this killed off another monojet. "The observation of that event," he said as we walked through the damp rain to his office, "the W with a recoil jet, implies that you must have a Z with a recoil jet in the same energy range. The same production mechanism is active for both W's and Z's. Instead of having here a W recoiling against a jet, you can have a Z recoiling against a jet. And when the Z-zero chooses a neutrino-antineutrino decay mode, you have nothing else to see but a 40 GeV recoil jet. And this is what we would call a monojet. But it would clearly not be due to any unknown physics. It would clearly be due to the tail of known physics. That's the point."

Back at his office, Denegri ran through all the different ways you could get monojets from known physics. He pulled out picture after picture and chart after chart from his briefcase. All the time he was talking to himself and pacing about his office: "I have such events I can show you. But do I have a nice event of that sort? Do I have it here? Let me see. I'm going to miss everything. Do I have here a Megatek of a nice tau decay which also generates monojets? I think I have it somewhere. Where did I put my nice events of today? Ooooh, here it is."

He showed me plots of all the strange W and Z decays, all with jets, some large, some small. He pulled one plot of W events out of his desk. It was a square, in which he put W's with very energetic electrons and not much missing energy in one corner, and W's with less energetic electrons but large missing energy in the opposite corner. "Every single point here is equally probable," he said. Then, pointing to different corners, he explained: "If an event falls here, you call it electron-jet. If it falls here, you call it electron-neutrino jet. If it falls here, you call it monojet. You see the point? If it falls here, it is still a W, but it explains one or two monojet events."

"Now," he said, laughing, "I have to convince my colleagues. That's going to be harder. That's not a small order."

From his office we walked back to the scanning room. It was after midnight. An hour earlier, Denegri had left his young Saclay associates working on the Megatek. Now they were gone.

"So where have they gone?" he said, as he sat down to work. "Ah, these young people, they don't have the energy to do physics."

12.

DECEMBER, 1984:
THE TALE OF THE TAU

"Okay, folks. Here we are. Where is the victim? Where is the person to be slaughtered?"
  —Carlo Rubbia, beginning a UA1 physics meeting, December 4, 1984

It was one of those stories at which everyone laughed, even though they knew it could have happened to them. A woman physicist had been waiting for a couple of weeks to steal just a few minutes of Rubbia's time and discuss what she considered a crucial and highly important piece of physics. Rubbia also thought it was important, but he had been flying around the world, coming and going, and the physicist had just about given up hope.

Finally, one morning she gets a call from Rubbia. She picks up the phone and Rubbia says, "Okay, I have exactly twenty minutes to talk to you about your work." This is great, she thinks. She slams down the phone, runs full speed to Rubbia's office, making it there in about ten seconds, only to find that his door is locked. She turns to Rubbia's secretary and says, "Carlo's door is locked?"

"Yes," the secretary replies. "Carlo was calling from the airport in Zurich."

Meanwhile Rubbia has called back and is saying to his secretary, "What the hell is the matter with that woman? I tell her I can talk to her about her work, and she hangs up on me."

This was one of the problems of working in a collaboration as large as UA1. Theoretically, anyone interested in working on analysis could join one of the existing teams. Or if they preferred individual accomplishment, they could write a computer program that would run through the thousands of UA1 data tapes and sift out the events in which they were interested. They could then feed those events onto more tapes and take them home to Paris or Vienna or Riverside, and try to do some physics.

In practice, it was next to impossible. Rubbia refused to look at any work that he had not first blessed, and he tended to explode at

171

anyone who presented individual work in a meeting without first briefing him in private. But to present your analysis, you needed a physics meeting, and Rubbia might only schedule one every three months, especially if he was not overly excited about the topics. The man was simply not in town often enough to talk to all the physicists who had ideas for physics. Some resorted to waiting on the stairs in front of his office. Some asked his secretary to relay the message that they would like to talk when he was in town, and then waited for their phone to ring. Some just did the work anyway, and waited for the explosion.

Rubbia simply liked to keep the analysis under his control—and that meant at CERN. Some groups, like Anne Kernan's Riverside people, insisted on working on analysis at home, but they picked relatively unimportant topics. Others, like the Annecy physicists, stayed right on top of the physics, and were in there pulling out important results, but were close enough to Geneva to commute daily when the pace picked up.

Then there was what Rubbia thought of as the Paris contingent —from Collège de France and Saclay—who seemed perennially unable to come to an agreement with Rubbia as to how, and how much, they could work back in Paris. Like many of the French, they simply refused to see why they should leave their beloved capital, let alone spend more than the absolute minimum of time in a suburb of Geneva, in order to find job satisfaction. They also had this annoying tendency of coming up with innovative ideas before the CERN group thought of them. Unfortunately for them, however, if they managed to convince Rubbia that an idea was worth pursuing, he then wanted it to move at his pace and under his eyes. Which meant moving it full time back to CERN. It was a small war of attrition. If a physicist preferred not to live at CERN, Rubbia might suggest that one of his local people attack the idea. Then a few other local physicists might join up, putting in their usual twenty hours a day on the CERN computers, which were faster and more powerful than any in Paris. By the next time the recalcitrant physicist returned to CERN, he would find himself three months behind and fading fast.

The two main culprits in this Paris contingent were Denegri and Aurore Savoy-Navarro. Denegri claimed that he always tried to have his latest analysis nearly complete before he broke it to the physicists at CERN. But he had not had much luck with the results.

Savoy-Navarro had the added burden that she was a woman and

believed that when she came up with a good idea, she had as much right to pursue it as anyone. And she would not back down. Sometimes this angered Rubbia, sometimes it amused him. Rubbia was not an extreme chauvinist; he might get caught calling his female physicists "girls" now and then, but basically, if they did good work, he recognized it as he would if it had come from a man. But he did seem to expect them to look up to him as a kind of beneficent father-figure, and if they did not, as Savoy-Navarro did not, then they had troubles.

These troubles of Savoy-Navarro's peaked when she resolved, with her Parisian colleagues, to prove that UA1 had seen the W decaying into a tau and neutrino. The implications of such a proof had become critical. The collaboration had to prove that they could recognize the W to tau decay when they saw it in order to argue that the monojets were not the product of those decays and were, therefore, the product of a more unusual effect; for instance, the presence of a supersymmetric particle.

Rubbia had said over and over that he was convinced that at least some of the monojets were not W to tau decay, but every time a UA1 physicist gave a talk on the subject, someone in the audience would say, "Are you guys sure that you're not just seeing taus?" It was obvious they needed to show the W to tau. "An absolutely fundamental problem" was what Rubbia called it. Unfortunately, absolutely fundamental problems were the ones that Rubbia tended to give to his disciples as high-priority items. And in early December, the tau analysis was in danger of becoming such a donation.

Aurora Savoy-Navarro was in her late thirties, with brown eyes and short straight brown hair. Around CERN, she tended to wear a grim, determined expression, and, like everyone else, she looked a little jaundiced from too much work and too little sleep. Unlike most physicists, she seemed to be aware of the tides of fashion in the world, and her dress reflected this in a subtle way.

Savoy-Navarro had become interested in physics early. Her father had wanted to be a physicist, but he was driven out of Spain during the revolution and never had the chance to finish his studies. Savoy-Navarro was then fortunate enough to grow up in France, where a woman could have a female physicist—Marie Curie—as a role model. The first time Savoy-Navarro read of Curie, she was hooked.

She studied physics in Strasbourg, then went off to CERN to do her doctoral work, studying the hardware of what she called "a horrible piece of detector," but which she turned into a working apparatus. "They gave me that," she said, "like saying, 'Try to do something if you can.' People at CERN do not educate you. You have to educate yourself. It can be very bad. It can be very good." In the late seventies she worked with Steinberger on his neutrino experiment—the "Is There a High-y Anomaly?" experiment—and by 1979, she was ready to go to work on pbarp.

Steinberger encouraged her to go with UA1 because the physics would be better than at UA2. "I was hesitant because of what I knew about Carlo's personality," she told me. "The legend was that if you go with Carlo, you become essentially a slave. But I decided that if you don't behave like a slave, he won't treat you that way."

Savoy-Navarro started at UA1 on hardware and then, after a couple of years, made the mistake of engaging in an initiative that was not first sanctioned by Rubbia. It started in a roundabout way: In the winter of 1982 Dimitri Nanopoulos, the CERN theorist, broached to Charling Tao, a young Harvard physicist who had taken a post at Collège de France, and who knew Savoy-Navarro, the idea of a holding a supersymmetry workshop. Tao had just come off three years of hell working on the UA1 central detector, and was now looking for a chance to do some physics. This was still a year before the W discovery, but supersymmetry had become the hot topic among the European theorists and looked like an interesting concept to pursue.

Tao suggested to Rubbia that they might organize a workshop between experimentalists and theorists to discuss how supersymmetry could be found in experiments, particularly those like UA1. Rubbia rejected the idea as "rubbish" and suggested to his collaboration that they ignore it. Because Savoy-Navarro was, as Tao put it, "a person who never listens to any interdiction," and because she had the added benefit of a husband who was a supersymmetry theorist, she joined up. The workshop, which came to be known as the SUSY workshop, was held in April of 1982, with lectures by both theorists and experimentalists from Europe and the States.

Rubbia's response throughout was cold. Immediately after the workshop, he shunted Savoy-Navarro off onto drudge work on the high-voltage system for the central detector. Tao, after those three years on the central detector, took a six-month leave of absence.

* * *

From the SUSY workshop Savoy-Navarro had an idea of how super-symmetric particles might show up in the UA1 detector, and in the summer of 1983, after the work on the W and Z had slacked off, she began to search for signs of supersymmetric particles in the data. With two other UA1 physicists, Savoy-Navarro would work after midnight at CERN sifting through the data tapes for events with the signature of SUSY particles. By August they had some interesting preliminary results, but by then Rubbia had decided to take over the bandwagon.

A month earlier, Rubbia had left for a physics conference in England. His opinion on supersymmetry at that time was of the Glashow school of thought—somewhere between cynical and to-tally close-minded. Rubbia returned from England raving about the theory. What brought him around was a mystery. One suggestion was that Weinberg sold him on supersymmetry, and that Rubbia had great and justifiable respect for Weinberg's judgment. Another was that Rubbia realized at Brighton that two prestigious Italian theorists had switched to SUSY, and he had always had great faith in his countrymen. When in early 1985 I asked Rubbia what had turned him into a supersymmetrist, he replied obtusely, "It is obvi-ous to me that one of the basic diseases among Nobel Prize winners is that they don't believe they have the right to change their mind any longer. This is something you have to be very careful about."

One way or the other, Rubbia had fallen for the idea of super-symmetry. From that moment on, Savoy-Navarro had problems. After she presented her work in an August, 1983, meeting, Rubbia called her and her collaborators into his office and scolded them for working independently without keeping him informed. Then he sug-gested to Jean-Pierre Revol that he look into the SUSY question, and do it under Rubbia's eyes at CERN. Revol had actually looked for SUSY particles in Ting's detector at DESY years before, and had spent part of the summer in a summer school in Corsica studying supersymmetry with Nanopoulos.

Rubbia tried to force Revol and Savoy-Navarro to work together, but the attempt failed. Both thought they should lead the analysis, and both could be stubborn and domineering. On top of that, Savoy-Navarro was trying to split her time between the physics and her two young children, and the burden was wearing her down. The partnership foundered. At one point Revol and Savoy-Navarro were turning to different theorists for help, because they would not trust a theorist who might already have consulted with the other.

By the end of 1983, Savoy-Navarro had moved permanently to Paris—where her theorist husband had taken a position—and was pushing the analysis from there, but without either the manpower or computing power that was available at CERN. Now, with the help of Denegri, Charling Tao, and a few post-docs and grad students, she was ostensibly no longer looking for SUSY particles; instead, she was trying to establish that UA1 had seen the W decay into the tau and neutrino. The tau analysis was the perfect answer to Savoy-Navarro's problems—although in the first instance Denegri and Tao had to convince her of that.

The motivation was threefold. Firstly, because the W to tau was the most likely background to monojets, it would have to be understood before any claims for SUSY discoveries could be made. At the same time, any possible monojets that were actually due to SUSY would also come out in the tau analysis. Secondly, as Savoy-Navarro said, "To get a bunch of people interested, you cannot arrive some place new and say, let's look for supersymmetry. I would look a little bit crazy. So saying let's look for the tau was a pretense for SUSY." And thirdly, the tau analysis was still not sufficiently glamorous at that time for Rubbia to give it to his CERN-based physicists and let the juggernaut take it over. It was safe.

Savoy-Navarro presented her first results at a collaboration meeting in London in January, 1984, as did Revol. The two presented similar results—the most spectacular aspect of them being the three or four mysterious events that later became known as monojets. And then the juggernaut did take over. Three months later, when Rubbia claimed he had discovered new physics at the workshop in Bern, he used the analysis of his group at CERN—Revol, and by then Rohlf and others.

The CERN juggernaut may have accelerated the pace of the analysis, but Savoy-Navarro—or for that matter any physicist—was motivated by more than just the desire to contribute to the pursuit of pure knowledge. Ego was another factor, manifesting itself in what Denegri called exhibitionism. "Intellectual exhibitionism," he explained. "You want to be first on something and you'd like that to be known. That's part of the game."

Then there was the unavoidable game of publish or perish. Even though it was hard to judge who was behind a particular piece of physics when it came from such an enormous collaboration, the physics community always seemed to learn. Usually, it was the physicist who gave the talks (or, at UA1, the physicist who gave the talk after Rubbia gave it first). Presenting work at a conference was one of the pearls that Rubbia could dole out to physicists who played his game, or withhold from those who didn't.

The knowledge of who did the work would seep out, and once it did, it could buy the next step up the career ladder. In the States, this could mean either a hot assistant professorship or a prestigious tenured position. In France, the job situation was more complicated. Within the French academic bureaucracy every major step on the ladder was divided further into eight levels. To move up those sublevels was not difficult; to jump from one step to the next was almost impossible. "The situation," Denegri explained, "was so difficult that you really had to do something spectacular." Although Savoy-Navarro was pessimistic about how far she could go ("The way things are going in France," she told me, "it's not your scientific work they value. These things are decided usually on lots of internal politics"), she had this habit of fighting. If the monojets were new physics, the CERN contingent and Rubbia's disciples were going to get the credit one way or the other. So Savoy-Navarro and the Paris contingent continued to work on distinguishing the signal for the tau, something the disciples considered, at the time, nearly impossible.

By late summer, 1984, the Parisian group had proved, or so they thought, that they could establish the signal in the 1983 data; they could prove they had seen the decay of W's into taus. By mid-September, Savoy-Navarro and her colleagues had drawn up a report, intending to distribute it to the UA1 physicists. First, Savoy-Navarro gave a copy to Rubbia. "Carlo said this should not be distributed like that," she told me. "He asked us to come to CERN, and said, this is interesting work and you should write a publication on it. He was very positive."

And then Rubbia won the Nobel Prize. He virtually disappeared for two months while, as he put it, he went "flying around from television camera to television camera." In late November, he and Savoy-Navarro finally distributed a first draft of a journal article on the W to tau analysis with a note signed by the two of them saying that they believed the paper to be worthy of publication.

The situation now became a bit uncivilized. By this time, the tau was a hot issue. The monojets seemed to be reappearing in the new data; they even had a few dijets, as the supersymmetrists' theories predicted. UA1 now had to prove they could differentiate the tau from the monojets, which they were claiming had too much energy to be taus.

In a note attached to the draft of their paper, Rubbia and Savoy-Navarro had requested written comments so that they could take these into account on the next draft. This was only the second paper from the collaboration not written directly by Rubbia—the first had been on a comparatively minor piece of analysis done by the Riverside group—and it was open to potshots from the collaboration. Physicists who had admitted that the monojet paper had been virtually unreadable, but had been given no time to voice their opinions, now acted like boxing fans who, having sat through an evening of unsatisfying bouts, suddenly find themselves invited to hurl tomatoes at the fighters.

The most immediately evident flaw in the paper was that Savoy-Navarro's English was not quite fluent. Rubbia's own English was a bit fractured, but he usually had his papers rewritten by one of his disciples for publication in scientific journals. Savoy-Navarro had had no such help, and some of the physicists jumped on the English while ignoring the physics. Others, however, jumped on the actual data, which they claimed were inconclusive. The background was a swamp of tau-like phenomona, and the analysis barely raised the tau signal above it. On the other hand, in the present 1984 run, UA1 had installed a new trigger, the missing-energy trigger, and with that, plus twice as much data, they had come up with what appeared to be obvious W to tau decays. These seemed to need little more analysis than simply looking at the reconstruction on the Megatek. ("You talk to Rohlf," Rubbia told me one afternoon, "and he will show you very nice examples of W to tau.")

Most of the criticisms directed at the tau paper were couched in the form of helpful suggestions as to how to further differentiate the signal from the background. But the remarks from Carlo's Disciples seemed less constructive. Rohlf was typically brash in his note to Savoy-Navarro. "The objective of this paper," he wrote, "is to make one physics point, namely, that tau decays of the W have been observed. However, on this point the authors fail completely. . . . To further insult the reader, the English is atrocious. On the positive side, I believe that the first sentence of the introduction can probably be salvaged for future considerations." Another of Rubbia's disciples simply wrote that he considered the paper uneditable and, therefore, unpublishable.

Meanwhile Rubbia had returned from a weeklong trip, and upon reading both of these letters, informed me that the paper would have to be rewritten by "you know who." Then he supposedly told one of his disciples that it was "good the young people were doing this because that way he didn't have to."

The physics meeting to discuss the tau work was scheduled for December 4; it shaped up as a confrontation between Savoy-Navarro and the Paris contingent, who wanted to publish on the 1983 data, although admitting it was not completely conclusive, and Rubbia's disciples and the CERN contingent, who had control of the 1984 data and knew they could easily show the W to tau decay. Savoy-Navarro would present the work, she told me, because none of her collaborators particularly wanted the job. Rubbia intended to be diplomatic, but his sympathies seemed clear.

The three o'clock meeting drew the biggest crowd of the year, and it hummed with chatter. Rubbia arrived a quarter of an hour late. But nobody was going to start this meeting without him; nobody even knew how. Rubbia was wearing a new gray pinstripe suit and a haircut for the trip to Stockholm the following week. ("Going to pick up a check," he said.) He was in good humor.

"Okay, folks," he announced. "Here we are. Where is Jeanne d'Arc?" He peered into the crowd like a nightclub magician searching for a volunteer for his knife-throwing act. He found Savoy-Navarro, who took center stage while disclaiming any leanings toward martyrdom.

"All right," Rubbia said. "Bong! First round."

For the next half hour Savoy-Navarro patiently explained the Parisian analysis point by point. Rubbia interrupted every few minutes, never totally relinquishing the stage. When Savoy-Navarro finally hushed him, he turned a few minutes later to joke with the physicists sitting to either side of him. Savoy-Navarro gamely tried to concentrate. When she had finished, she walked to the corner of the room and slumped against the wall as Rubbia took over.

"It's quite clear," he began, looking for some reason directly at Denegri, "that the understanding of the tau decay mode is important. We can't believe the monojets until we believe the tau."

Denegri nodded, and Rubbia lit a cigarette. He paced with one hand in his pocket.

"Now you people have done a lot of work," Rubbia continued, "and it's perfectly normal to see that effort put together in a document and we have discussed that several times, and of course the document is not perfect. There are certainly things that are not right, etc. . . ."

Rubbia sat on the table in front of the room, removed his jacket, and asked for questions and comments. With each response, he made a suggestion of his own: "Let's see if we have any opinion on this," he would say, "or if you like you can go directly to Jim Rohlf 's talk." And then a few minutes later, "I suggest if there are no more questions, that we hear from Jim."

Rubbia was either strapped for time, or he simply wanted the audience to see Rohlf 's show before wasting time on more discussion, since Rohlf had the 1984 data. Finally, Rohlf took the stage. His hair was mussed; and his eyes, as usual, were half-closed and puffy from too much work.

He flashed one transparency after another. Very fast, with no analysis, just letting the Megatek pictures speak for themselves. Nothing but extraordinary events. He started with monojet events that he thought were certainly not taus. Everything over 40 GeV missing energy—the cut-off point, beyond which they had claimed that it would be impossible to generate a tau. On a few Rohlf said, "I would claim that this cannot be a tau." For the others, he had no doubt. Then he moved to the ones that could be taus. And he was just as fast, just as certain. "Very characteristic of tau," he said, showing the Megatek pictures. "Of course, it's nice to have more than one. And here's another. Notice the missing energy is all much lower than the monojet region: 15 to 30 GeV. Here's another. There's no chance that this is anything but W decaying into tau."

Rohlf had his sleeves rolled up. Rubbia was leaning back in the front row, with his hands behind his head, smiling. This was the kind of show he liked.

Rohlf finished up and took his seat.

Rubbia then made his suggestion.

"Should we join efforts with '83 or '84," he said, "or shut out '83 and follow up with '84? My gut feeling is '84 looks extremely appetizing and we should use it. Yes or no?"

"We have also looked through the '84 data," Savoy-Navarro insisted. "Of course, it's absolutely sure we've got much better data, but the point is that . . ."

"I'm talking about having a group formed of people," Rubbia interrupted. "All of you people and maybe some more. Get a task force which could squeeze out the '84 data."

Rubbia's argument suddenly became evident. There was no

longer any question of going with the 1983 tau paper immediately. Instead, he wanted to wait just a few months and write a conclusive paper from the 1984 data. Although several in the collaboration argued that a good piece of physics from the new data would take at least six months, Rubbia and the disciples disagreed.*

Savoy-Navarro saw the juggernaut beginning to roll. Any task force would be CERN-based, and if she wanted to be in on it, she would be at best one of many. But in this case the majority of UA1 was with Rubbia.

Savoy-Navarro and the Paris contingent were stymied. They tried to argue that their analysis was no less muddied than that presented in the monojet paper—"If we can't show that something we know to exist, exists," said one mystically, "I don't know how we can show that something not known to exist, exists." But even though they had sympathizers on that point, it was not an argument that would ever work. Savoy-Navarro argued justifiably that if Rubbia had been at CERN instead of talking to television crews about the Nobel Prize for two months, a paper on the tau could have been out before they had had 1984 data to contest it.

But Rubbia had preordained the outcome. When the meeting ended, one of the British physicists told me simply, "Very deftly handled." Another added, "There are no results that can come out of the laboratory other than CERN's."

As the other physicists left, Savoy-Navarro and the Paris contingent sat together in the middle of the room and argued about what to do now. The Paris contingent considered Rohlf's show a kind of antipropaganda. "As far as I saw it," said one, "he was saying in a few words, all you've done is bullshit. That is not very constructive."

Chiding Tao insisted she was not angry but still seemed at least hurt. "Carlo made one comment that I overheard," she told me. "He said it's the first time he saw so many criticisms of a paper. He shouldn't forget it's the first time that the paper doesn't come from him."

Savoy-Navarro took the TGV back to Paris and went immediately back to work. The collaboration had agreed to meet again in January, 1985, to discuss the future of the tau paper. She wanted another

*In fact, a year later UA1 had yet to publish a W to tau paper, for reasons that will become obvious later in the book.

draft finished by then, with the constructive criticisms from the collaboration taken into account.

Toward the end of December, I asked Savoy-Navarro what would happen now. It had been two weeks since the meeting. "We are going to distribute the version we have rewritten," she told me. "People have to decide soon whether to release this paper. In three months it's clear that it will start to be out of date. Now we are one hundred percent sure that the world is round and tau exists. We think that the work is done and people have to realize that."

A month later, the question of whether to publish was brought to Rubbia's executive committee. Nine of the UA1 institutions voted for publication; three voted against. No one would say, or was told, which were the three. Because Rubbia would not let the paper be published without unanimity, it was shelved.*

If the tau meeting was particularly vitriolic, one of the reasons might have been that the collaboration was suffering one of the worst collective cases of overtiredness in recent memory. The run was winding down to the Christmas break, but the data were still pouring in. The physicists had been through three months of flying to Geneva, working nights or round the clock, and flying back to their families and their classes and their students. And it showed. In the control room, once the beam was in, they tended to nod off and hope that nothing happened that needed attention. A box that contained change for the coffee machine was somehow always empty.

This fatigue was not helped by Rubbia's promise to the world in general that they would have the data analyzed by Christmas or shortly thereafter, as had been the case with the W. But they had twenty times more data than from the W run, and the closer they got to Christmas, the farther behind they fell.

I would walk into the Megatek room at midnight and find Pauss, Rohlf, Mohammadi, and a few others just coming back from dinner, getting ready for another five hours of work, and this either after or before their shifts at the control room. They all had bloodshot eyes and sallow complexions. They tended never to go anywhere without a cup of coffee in their hand, or at least a pack of Marlboros. And many of these people normally were nonsmokers.

---

*In late February, Rubbia requested that Savoy-Navarro report on the tau at the Proton-Antiproton Physics Workshop in St. Vincent, Italy. It was his way of compensating Savoy-Navarro for her work. The physicists who attended St. Vincent would tell me afterward that her talk was one of the high points of the conference.

Back in October, Felicitas Pauss had told me that the only time she went home to her apartment was for six hours a night to sleep. And even then she would have trouble falling asleep because she could not stop thinking about her work. Then she would wake up in the morning with the same thoughts intruding before she even realized she was conscious. She had suggested that what they were doing was destructive; that when she worked so hard, she became inefficient and began to hate her work.

Pauss was Austrian. Her father was a musician. She had played the piano and the violin but had chosen a career in science because she felt that she knew the routine of a musician and could find more creativity in physics. "In an orchestra," she told me, "you have to follow the conductor. You have to do what other people tell you. Now I have the feeling that there is no limit to my creativity. When I scan an event on the Megatek, nobody tells me what I'm doing with the event. I have to use my brain, my creativity."

Pauss had joined the experiment only after the detector had been finished, and had never worked on hardware, only on analysis. She had, in the process, become one of Rubbia's lieutenants. She had done so on the strength of her eighteen-hour days. Unfortunately for her, however, because she was very attractive, and because no matter how many hours she logged she always seemed to show up in extremely flattering outfits, some of her colleagues were suspicious of her proximity to Rubbia. They felt she had not been properly baptised in the sweat and tears of building that 2,000-ton piece of history.

When I interviewed Pauss, she suggested I should ask her directly if she felt discriminated against. When I did, she told me that the disadvantages of being a woman in physics were that some male colleagues still had fixed opinions on how she must dress and behave, and that they were wary of any woman who trespassed on what they considered to be their territory, which Pauss obviously considered her right. "The advantage," she said, "is that if you go to a conference and there are over fifty talks and only two by women, people tend to remember you. I seem to know more people than other physicists do, without making any more effort."

Now she had Rubbia's ear, but paid for it in terms of her life outside of the lab, which by December was nonexistent. Her six hours at the apartment had been reduced to five on even a slow day. She would leave the Megatek at four or five in the morning and return to it by ten. Near the end of the run, she told me that she couldn't tell the difference between weekends and weekdays anymore. She had missed an appointment when she couldn't remember which day of the week it was and went on the wrong day. And she said that she dreaded all incoming data, because every time they scanned new events, it simply resulted in more work. Still, she was amazed at how much energy she had to keep working and working, well after she thought she would crack. Christmas was just around the corner, and she would be going home to Salzburg.

Would she take work home with her? I asked. Absolutely not.

This chronic exhaustion was exacerbated by the performance of the SPS, which had been creaking along as though it too couldn't wait for its vacation. The machine physicists had given up hope of getting the luminosity up to 500 inverse nanobarns, which had seemed earlier in the year to be a magical number, but within the realm of possibility, and were ready to consider themselves lucky if they got 400. The SPS crew had been using the machine frequently for what they called machine development, which meant experimenting with the accelerator itself. But every time they dickered with the beams, it took them a few days to get the machine back into top form.

The SPS people had their share of problems as well: one day it was a power cable that had grounding difficulties, resulting in the loss of the beam. Then it was a technician accidentally switching off a cooling pump, again resulting in the loss of the beam. Then it was a power supply that dropped just a couple of minutes after the beams went in. This had never happened before. On December 17, Hans Hoffmann, in charge of technical affairs at UA1, told the accelerator physicists that if they hit 400 inverse nanobarns, he would buy champagne for all of them. Nothing was more demoralizing than not getting your physics.

Two days before the run ended, the mood momentarily lifted. While working on the Megatek, Pauss found a two-jet event which had a huge missing energy of 63 GeV. A beautiful event. They had scanned a little more than half the data they had taken—the express-line data, the stuff that the computers selected as potentially interesting. When I asked Rohlf whether this two-jet event was just what John Ellis and the other SUSY theorists had been looking for, he had his usual cocky answer. "Yeah," he said. "But don't tell him. He'll go nuts."

I went looking for Rubbia to ask him what he thought of the event, and found him in his office at ten o'clock at night. He had flown to London, given a lecture, and flown back the same day. Now he was walking into his secretary's office to check his answering machine and look in her desk for the pack of Gauloises she kept for moments like this. His shirt was untucked, his eyes were puffy, his face was lined and his hair mussed. He looked more exhausted than I had ever seen him. He walked out to squint at the television monitor above his office door, and pointed up to the inverse nano-barns. It had been a good day. "Look how nice it's working here," he said. "377. We'll get 400. It's something we want to have."

I asked him about the two-jet event, and he said, "It's really a fantastic thing." But he was more interested in telling me his other news. They could now prove, he said, that the results that UA2 had claimed earlier in the year might be new physics (the original events of W's with jets that had started the brouhaha the previous February) were, in fact, just the tail of the distribution. Rubbia had been given the latest UA1 analysis from Steve Geer, one of his lieutenants, and they now had enough statistics to prove that UA2 was wrong. The effect was no longer an effect.

In the previous run in 1983, Rubbia said, UA2 had thirty-five W's, of which four were the strange high-energy variety that might have been new physics. This year UA1 had two hundred W's, nearly seven times as many as UA2 had had the previous year. If UA2 had been correct, UA1 should have approximately twenty-eight of the high-energy W's with jets. But they had at most one. His conclusion was that UA2 had been screwed by statistics.

Rubbia explained this concept to me in another one of his metaphors. "There is always a numerator and a denominator in these things," he told me. "The numerator is the number of times you win, and the denominator is the number of times you try.

"You have the denominator, which is the background, and the numerator, which is the signal, and the signal is only a signal if it occurs at a high rate. So now the denominator keeps ticking away, higher and higher and higher, and the numerator stays put. So the overall signal keeps sinking down and eventually there is a bottom line, a place where you are in the mud. At this moment, they are diving into it. Those events they have are still there. You can see them; I can see them. But their significance is reduced.

"You could say UA2 is bullshit," he added. "But that's not true. What you say is UA2 had the effect early in the game. Therefore, within the margin of error, they overestimated their curves."

*   *   *

Early in the morning of December 20, the machine physicists, sitting in their control room in France, shut the SPS down for the year. They did not make it to 400 inverse nanobarns, peaking out at about 398. They sacrificed their champagne for six hours of machine development time the day before, running an experiment that would be important when they got around to building the next generation of accelerators in five years.

At UA1, the midnight to eight shift was working, with a few extras who had drifted in to watch the finale. Cline was on the muons that last night, writing letters to physicists inviting them to seminars, and looking as though he needed to get away to Barcelona for the holidays, as he was planning to do.

At five in the morning a film crew came in from the States. They had flown in to film Rubbia for a special on physics and the universe. They looked weary and jetlagged as they set up their equipment and peered around a bit nervously for something interesting to shoot. It was quiet. Since the night before, the monitor had read: "We hope to keep this until 0600, end of run. In case of loss, refill up to midnight. Beam will be dumped 0600 sharp." At twenty before six, the SPS crew called to warn UA1 that they would drop the beam on time. At five of six, the UA1 people shut down the central detector. At 6:05, five minutes behind schedule, they dumped the beams. In the control room the only indication that the run had ended was that the two lines on the antiquated oscilloscope that tracked the proton and antiproton beams went flat in the blink of an eye. That was it: no more collisions, no more beam. They had all the physics they were going to get for the year. There was some scattered laughter, a few jokes, a few sighs of relief. A cameraman from the film crew asked if that was all there was.

13.

JANUARY, 1985: THE UNIVERSE ACCORDING TO UA1

"We are only going up to the Sinai to get the tablets. We have no right to write the commandments." —Carlo Rubbia, on whether supersymmetry is fact

Over the holidays snow had fallen and it was cold. In the city the temperature dropped to 22 degrees below zero in the wind, the coldest day in over twenty years. The wind gusted at thirty miles an hour, whipping the water out of Lake Geneva and over the retaining walls, where it froze into a still life of a waterfall. Out near CERN, the wind drifted the snow off the fields toward the laboratory, and left it two feet deep on the highway into town.

The laboratory was deserted. The cantina was open but empty. The UA2 corridors were empty; the theory corridors were empty; the LEP corridors were empty. At UA1, the usual weekend crew was at work.

Rubbia was in. He had just returned from a vacation in Africa and was telling tales of his safari. He seemed relaxed. Rohlf was in, working with Pauss on the Megatek. Rohlf had spent Christmas back home in Minnesota. He came in to CERN looking like he'd been hit by a snowplow, his hair disheveled and covered with snow, his blue overcoat hanging lopsided, and his Marlboros dripping wet. Pauss had spent the holidays with her family in Austria and was back with a box of home-baked Christmas cookies for the physicists. It was so cold in the Megatek room that they worked in their winter coats.

Mohammadi was in. He had been unable to get his car started in the snow. He had jump-started it and then, when he got to the office block, opened one of the huge garage doors in the back, driven his car inside, and parked like it was a prize on a television game show.

Mohammadi was preparing a talk for the winter workshop in Aspen. It would be his first conference talk. He was still a graduate student, and he would be discussing the hottest subject in physics. It would also be the first time that American physicists were exposed to the data from the new run in any conclusive manner. Rubbia had shown some of the flashier events of the new year in Sante Fe in November, and had talked about supersymmetry in a roundabout way, so that some of the physicists took him to mean that he was claiming he had discovered it. But it was Mohammadi who would get into the data for the first time in any detail, even though only half of it had been analyzed. He was nervous.

Before the break, he had covered some of his work in a meeting with Cline and the Wisconsin/Harvard crew. As Mohammadi showed his transparencies, Cline would interrupt him every few minutes: "Check this," he would say. "Check that . . . let's see this compared to that." Mohammadi had looked a little exasperated. Afterward, he had told me that Cline seemed to be overanxious about the new physics. "He is really trying his best just to kill the thing," Mohammadi had said, "rather than to check out other possibilities. That in a way is the easy way out. Then you don't have to try to explain it."

Mohammadi was not as pessimistic as Cline, but still he wished that Rubbia would back off a little and give the collaboration time to do the analysis thoroughly. He felt that they had jumped the gun the previous spring when they had published the monojet paper, and he was a little fearful that they were doing so again.

Mohammadi was in a tight position. The official UA1 stand on the monojets, as decreed by Rubbia, was that they were new physics. But Mohammadi knew that quite a few skeptics would be seated in the audience and they would love nothing more than to jump all over a UA1 physicist, shouting "High-y anomaly!" In fact, Cline had been trying to prepare Mohammadi for such attacks, which he himself had experienced too many times.

"I can bullshit forever in a meeting with Dave and other people," Mohammadi told me. "I'm not afraid of making mistakes because, after all, I'm here to learn. In a big conference, you're not supposed to do that, even if they know you're a student. But it's the nature of this talk that I'm nervous about, because this is the kind of talk in which people come to tear you apart. Right now, I don't know how much I should defend this position of new physics. It's clearly been said in the paper and I know I probably have to stick with that. But at the same time, if I come out really confirming this stuff, then people take the liberty of asking any single possible question they can to make sure I'm not bullshitting them."

A few days before the holidays I had found Mohammadi in his office picking the color of ink he would use on his transparencies. He had his first page carefully penned out. The introduction was headed "EXPERIMENTAL RESULTS" in nice clean capital letters. The first page had to look good, he told me. But that's as far as he had gotten.

He had spent the vacation with Genevoise friends skiing in the Alps. He had not thought much about physics. Now that he was back, he was feeling a little guilty about his vacation, and under pressure. On Saturday afternoon he went to see Rubbia, who told him to finish his talk before he would discuss it. They had plenty of time, Rubbia said. Mohammadi was scheduled to leave the following Tuesday morning for Aspen.

By Monday night Mohammadi was ready to show Rubbia his transparencies. Rohlf would join him. Rohlf was also planning to talk at Aspen, and he wanted to brief Rubbia first on the 1984 data and then let Mohammadi give him the transparencies from his own talk.

At ten-thirty, the two caught up with Rubbia. They went down to the conference room, where Rohlf turned off the lights, turned on the projector, and began his briefing with Rubbia and Mohammadi looking on. Rohlf was wearing his tweed jacket, Rubbia a white shirt, tie undone, and Mohammadi an old rugby shirt and two days' growth of beard. Nobody else was in the building. Outside, the snow fell.

Rohlf first showed a chart of all the missing-energy events. This chart had been the key piece of evidence in the data last year, and they had added the new data to it this year. The monojet and dijet events were set down as little squares and circles, plotted according to the amount of missing energy. The chart showed the tail of the distribution: If the missing energy in an event was small, it was placed to the left of the chart; if the missing energy was large, it was placed to the right. There were a lot of dots on the left side of the chart, decreasing in number toward the right. Slashing vertically through the chart was what they called the four sigma cut, which signified roughly where they had decided the tail of the standard model faded off into insignificance. To the left of this line were the events in which the missing energy was unimpressive. These were most likely taus. To the right were those in which the missing energy was mysteriously large.

Rohlf walked to the screen, and with his cigarette firmly tucked between his first and second fingers, pointed out the four sigma cut, and where the new dijets and monojets fell.

Then he put up two more charts. One showed the estimated background from gluon jets faking monojets, and the other was from Mohammadi's Monte Carlo. Both were very preliminary and seemed to show that these standard sources could not account for all the monojets and dijets.

Rubbia stood quietly. Finally, he walked up to Rohlf and waved his fingers in front of him, saying, "You have some smoking material?" He took a cigarette and went to work.

He described his reasons for believing that they were dealing with a breakthrough, why the monojets and dijets were inexplicable by the standard model. As he spoke, he became more and more frenetic. He paced back and forth, waving his hands, scratching his chest under his shirt.

"Listen," Rubbia said. "There are three questions you can ask yourself. One, are we dealing with a phenomenon produced by a problem with the instrument? You say no. Two, is this due to trivial phenomena? Answer is no. You show that distribution." He pointed to one of Rohlf's curves, which seemed to show that the number of W to tau decays expected from Mohammadi's Monte Carlo matched the current rate of known tau decays from the data. "Three," Rubbia continued, "is the Z recoiling against a jet or jets? You say no. The next question you can ask is, are these events some kind of exotic decay of the Z? Answer to that is also no, because the distribution would have to center around a mass and there is no evidence for that. . . ."

Rubbia was doing all the talking. This was his show. Rohlf sat next to the projector with one leg folded under him. He said nothing but watched Rubbia intently. Mohammadi looked a little dazed.

"We are pushed more and more by exclusion," Rubbia said, "to the conclusion that something like a supersymmetric particle is being emitted. This is an important statement you can make."

He stopped pacing and looked at Mohammadi. "Is this the first time you've heard about this?" he asked.

Mohammadi shook his head, but Rubbia wasn't paying attention. He walked to the board, then back to Mohammadi, and then back again to the board, standing with his right hand on his waist, gesturing with his left hand and his cigarette. "You don't understand this," he said. "If this is the first time you've heard this, you cannot talk about this. They will shoot you down in flames.

"Let's go through your stuff," Rubbia said to Mohammadi. The two leaned over the table and began paging through Mohammadi's transparencies. After a minute Rubbia sat down and leafed quickly through the charts and diagrams. He sat up straight with his stomach against the table, saying with each new transparency, "You go through this, you talk about the apparatus, you talk about this, this." Then he stopped. He asked Mohammadi where the dijets were, and Mohammadi rifled through his stack, finding the relevant chart toward the back. He handed it to Rubbia, who looked at it briefly and then laid it down on the others.

"This is a piece of crap," he said, chopping his hand down on the transparencies. "The multijets aren't described right." He stood and began pacing. Then he dictated Mohammadi's game plan. He wrote the key points on the board, as though he were a football coach telling his quarterback what plays he wanted run in the next game.

"You cover the following three things from the 1984 run," he said. "One, presence of monojets. Confirmation. Number two, dijets have appeared. Number three, the test of whether we are really doing something right is that we can identify the tau decays, which are the cheapest form of monojets. Period. Then you stop there. Then Rohlf can take over and summarize these conclusions and get into the business of interpretation. It is quite clear if you give an experimental talk you cannot do more than that."

Rubbia walked back to the transparencies and leafed through them once again. Finally, he walked around to the far side of the table, sat down, and stretched his legs.

"It's a fine talk," Rubbia said now. "It's an excellent talk. For the tau, you really know what you're doing. It's your thesis. So tomorrow morning we meet at eight-thirty and go over it. Now I go to sleep and you go to work. That's fair."

The briefing wound down. It was close to midnight. The snow was still falling. Rubbia lightly kicked a few chairs. He was smoking, smiling; his blue eyes were glowing. Rohlf told Rubbia he was going to Santa Barbara in a few days.

"What conference?" Rubbia asked.

Rohlf hesitated. "No conference. A lecture. A seminar."

"What's in Santa Barbara?"

"The Institute of Theoretical Physics."

"Theorists?" Rubbia said. "You are going to get in trouble. I can tell."

* * *

A few minutes later, Mohammadi stumbled upstairs. He decided that he would postpone his departure for Aspen. He would finish the talk that night, and see Rubbia in the morning. Then he would do some scanning, pack leisurely, and hope he could get a flight out on Wednesday.

He met Rubbia at eight-thirty the next morning. Rubbia was just about to leave for the airport and he had expected Mohammadi's transparencies to be waiting for him. "Where the hell are your transparencies?" he yelled. Mohammadi ran back to his office, grabbed the material, and then rode with Rubbia to the airport. Rubbia's secretary drove as Rubbia paged through the talk and asked a few questions, made a few comments. "He was a bit pissed off about the whole thing," Mohammadi said, "because he was late for his flight. I think he was happy with it. He said it's okay. That was the nice thing. I expected to hear him say something like this is a piece of shit."

The next day Mohammadi caught his flight and was off to Aspen.

Rubbia had laid out Mohammadi's talk point by point. Rohlf had a little more leeway. He was going to Santa Barbara, then on to a conference in Berkeley, and then to Aspen. Rubbia had, of course, set down his official position and Rohlf was Rubbia's chief lieuten-ant, his protégé. Nevertheless, before Rohlf left for Berkeley, he told me that Rubbia had been strangely nervous about what Rohlf was going to say. He had asked Rohlf about it and then hadn't given him time to answer. "Listen," he told Rohlf, "I don't want to know."

Rohlf, unlike Cline, firmly believed the monojets were new phys-ics. He had said so at the conference in Leipzig, the previous August, and his faith had yet to be shaken. ("Jim's being too emotional about this," Cline said to me one night. "He gave his talk at Leipzig on the new physics, and his ass is on the line.") So Rohlf was going to interpret the data as he saw it. He was going to say again that it was new physics.

"The conclusions of the talk are actually what Carlo outlined," Rohlf told me the night before he left for the States. Then he went through his transparencies and explained what came before the conclusions. "I have two talks which I have to separate: what we know and what we don't know. I spend most of my talk on missing energy. I'm going to say that the missing-energy events are confirmed and the events are real. They're not some artifact of the apparatus such as beam halo or some malfunction. Then I want to go into conventional physics backgrounds that we know about. . . . We're handling all the things that we know about. That's not to say that we've handled everything. We're handling everything we can think of."

Jim Rohlf was in an interesting position. He was a hot young physicist and in some ways the monojets were now his baby as much as Rubbia's. Jean-Pierre Revol and Aurore Savoy-Navarro may have started the work on the missing energy, but it was Rohlf who was the present heir to Rubbia's empire, and who, as an American, was invited to give the talks in the States.

Rohlf was in his fourth year as an assistant professor at Harvard. The position was good for eight. Then the university would either let him go or give him tenure. But Harvard had a reputation, at least among the young physicists, for never offering tenure to its own people, opting instead for established physicists who had already made their name elsewhere. (Rohlf described it thus: "Even if you're very good, you can still get flushed down the toilet.")

Rohlf took a relaxed view of whether or not he would be offered a permanent position. "This decision gets made when I decide to leave," he told me. We were sitting in his office, passing a bottle of Thai whiskey back and forth.

"If you look around," he said, "there aren't that many decent physicists. It's only a game and only a matter of style as to where you end up getting a job. A lot of people think that Harvard is a horrible place, for instance, because the people tend to be very politically oriented and very superficial in their physics. They deal with dollars; they will buy you physics. And, of course, they end up working on good projects. But they don't really contribute. They're just financial partners. That's the reputation that the place has, and the big exception is Rubbia. He's the poor kid. The Italian. They're really lucky to have Rubbia. Really lucky."

When I asked him which schools had better reputations, he listed MIT, Caltech, Berkeley, Cornell, Princeton, and Stanford.

"I have sort of the same philosophy as Carlo has on this thing," he told me. "The guy has a tremendous amount of enthusiasm and a tremendous amount of physics ability and a tremendous amount of power. So I'm trying to make use of that while I can, and when I leave, I leave."

When I suggested that Rubbia was also reputed to have been an electronics genius since his high school days and that this had brought him as far as his enthusiasm and his politicking, Rohlf just laughed. "The last time I saw him with a soldering iron in his hand," he said, "he was lighting a cigarette. Okay?"

Then he got serious, or more so: "You know why he's a genius? It's because he takes the technology from one continent, as I am doing now with these two EPROMS, Erasable Programmable Read Only Memories . . ." He opened his briefcase and showed me two little computer chips in a plastic bag. They were going back to the States with him for a new superfast electronics system being built there for UA1. "And he takes them from continent to continent to use them as necessary. So you just have to know how the parts work. You know what I mean?"

Rohlf planned to stay with UA1 for a few more years, and then, once he had tenure, lead his own group in one of the huge collaborations that would run the experiments for the SSC or the LHC in the nineties. His ultimate goal, like that of most physicists, was to be spokesman for an entire collaboration, the way Rubbia was for UA1. This was ambitious because it meant running a team of several hundred physicists, and Rohlf had not yet proved that he had that kind of leadership capability. But he had begun on the path that could lead to such a position.

Many of the more prestigious physicists were already impressed by Rohlf's brains, his opinionated style, and his affability. With Glashow, for instance, he would walk into his office and put his feet up and the two would bullshit about physics and personalities for hours. In a field where a large percentage of the upper echelon were either extraordinarily serious or somewhat socially stunted, Rohlf's forthright Huckleberry Finn demeanor was refreshing. He could alienate his peers relatively quickly if he chose, but he knew when he was doing it, and he rarely did it to those he respected.

For the past year, Rubbia had been using Rohlf as a junior partner in a deal to buy $3 million worth of non-weapons-grade uranium from the U.S. government and install it in UA1. Rohlf had started to learn the ropes of running a huge enterprise as well as of becoming tight with the men at the DOE.

The uranium deal had begun two years earlier. Rubbia had decided that depleted uranium would be the perfect material to upgrade the UA1 calorimeters, which had been designed in 1977. "Back then," Rubbia explained, "many of the phenomena that we now have around were wishful thoughts. Jets, for instance, did not

exist; just the products of theorists' minds." The uranium concept
was developed by a physicist at the ISR in the seventies, but had
never gone anywhere. Essentially it meant replacing the lead in the
calorimeters and gondolas with uranium. And if it really worked, it
would help UA1 measure the energy in jets with much greater accuracy. It was one of Rubbia's grand schemes to stay the frontrunner
in physics through the next six years, even as the SLAC Linear
Collider and the Tevatron were usurping the accelerator's lead.
"Since we will have an inferior machine," he told me, "we've got to
have a superior detector."

The uranium would have to be in and working by 1987, when
another of Rubbia's grand schemes would be finished, a second
antiproton accumulator ring that would increase tenfold the number
of antiprotons available to them. Then UA1 would have two years
of running before LEP came on.

They needed 300 tons of depleted uranium, what was left over
after the weapons-grade material had been separated out of uranium ore. Rohlf and Rubbia had been meeting with the DOE about
once a month for the past two years. It was only in November that
they finally had a rough estimate of $3 million for the price of the
uranium. In late December, 1984, they received the first ton for
testing. Getting the rest of the uranium out of the States would be
easy—theoretically it was on loan, and the government could ask
for it back any time. "Of course," said Rohlf, "they don't want it or
they would never lend it to us to begin with."

Getting it all into Switzerland would be a little harder. "We're
going to keep it low key," Rohlf told me. "We're not going to tell
anybody about it. The worst part of the shipping is the part within
Europe. The airplane is not going to land in Geneva. It can't go on
a passenger plane, and for that reason it can't go on Swiss Air,
because Swiss Air flights are all part-cargo, part-passenger. It's
going to land in either Germany or France, and then it's going to
have to be shipped by truck."

In the process of all this, Rohlf was learning as much about the
bureaucracy of physics as about the physics itself. "We had to go
talk to bureaucrat after bureaucrat," he told me. "We had to find the
uranium. Nobody really knew how much it cost. We knew it would
be up there somewhere, but we figured that after the W and the Z
were discovered, we had pretty good bargaining power. So first we
went to the DOE and told them we want money, and they said they
didn't have any money for us. Then they said they'd give us some

money but not enough money. By then we knew who had uranium. It was the United States government, as a by-product of the defense weapons industry. The first job was to convince the DOE that the cost of the thing was not astronomical, and astronomical would be considered something approaching the cost of the CDF,* which, depending on who you talk to, is on the order of $50 million. Then we just had to really convince ourselves that it's a device that can do some physics. Those guys at the DOE are real simple-minded. All they want to know is how much does it cost. If you make some reasonable case they trust your judgment."†

Rohlf had been spending a good amount of time with Rubbia running down both the uranium and money. The upshot was that Rohlf was as hot as a young physicist could possibly get. He was well respected by the upper strata of the physics community, and he had that all-important visibility in the back rooms in Washington. DOE bureaucrats who had never heard of Bernard Sadoulet or Mario Calvetti—without whom it is conceivable that UA1 would never have produced physics—were on a first-name basis with Jim Rohlf. He had already been offered a few tenure positions, but he was sitting back waiting for the best. If the new physics was the breakthrough he thought it to be, Rohlf would be able to write his own ticket.

On January 11, 1985, Mohammadi announced in Aspen the preliminary results from the autumn run. Rohlf then broke the news in Berkeley, and the following week he repeated it in Aspen. The monojets had been confirmed. Dijets had been found. Although the background calculations had not even begun yet, the message was that the new physics was still new physics.

Things moved fast. The same week that Rohlf spoke at Aspen, a team of theorists from Michigan sent a paper on supersymmetry to *Physical Review Letters,* citing information presented in Mohammadi's talk from the week before. At CERN, the theorists had not yet seen the data, but they'd heard about it. In December, Rubbia had flown to the Rutherford Laboratory in England, presented the limited data they had, and said that the new physics had apparently been confirmed.

*The first detector being built for the Fermilab Tevatron.

†"If a guy comes in with a proposal," Bernie Hildebrand of the DOE once told me, "and dresses in pink and walks on his hands, but if his physics gets laudatory reviews, I will close my eyes and say, go ahead, dress in pink, walk on your hands, and keep doing good physics."

This would have been a relatively innocuous talk if not for the fact that the next speaker on the program was CERN's John Ellis. Ellis had heard through the grapevine that UA1 had more monojets, but that was all. Then Rubbia showed his pictures of dijets, and Ellis decided that Christmas had come early. Ellis had been predicting that if supersymmetry was correct, dijets would show in UA1. When Ellis gave his talk, as he said, "I didn't have to change it very much. Since I had planned to say that you should see two-jet events, I made this point that indeed two-jet events are now seen, the rate is about right, and the characteristics of these events are roughly speaking what we'd expected."

Subsequently the British science journal *Nature* picked up on the two lectures and published a story that came only a hairsbreadth short of claiming that supersymmetry had been discovered.

These rumors had Rubbia in remarkable good humor. He was enjoying himself. "This kind of confusing situation is tremendously productive," he told me. "This is what we missed in the W and the Z; it was too simple to be fun." I wondered which he enjoyed more, doing the physics, or having the power to keep the entire physics community on tenterhooks. Rubbia would tell me that he thought the new physics had become "rock solid," and he would whisper this loudly to others in not-so-private conversations at the cantina. But still he did not publish, and his peers would tell you to look at what he published, not at what he said.

What he was saying now was extraordinary, however. He repeated to me what he had told Mohammadi: They might have discovered supersymmetry by default. Although he still believed it was good to be true. "Look," he said, "whether SUSY is there or not is neither Glashow's nor Weinberg's nor my choice. It's a fact that only nature has decided. We are just there to pick up the response."

Rubbia's disciples were considerably less certain than Rubbia himself. Rohlf would say that they had presented the data honestly, that it wasn't junk, and that they had argued that it was not old physics. It was easy to prove what it was not, however. "Yeah," he said, "we probably discovered new physics."

Most of the other disciples felt as Jean-Pierre Revol did: that it would take at least a few months of analysis before they could come out and say anything one way or the other. In late February there would be the Proton-Antiproton Workshop in St. Vincent, Italy. These workshops had become the traditional showcase for the new data from the collider. By then, perhaps, they would have a preliminary answer. Sure, Revol hoped it was new physics. When he heard that Rubbia was already claiming that it was, he told me that that was Rubbia's prerogative. "Maybe I don't have enough experience," he said. "Or maybe I like to see things with my eyes first and make sure I understand them."

The theorists at CERN were much less restrained. When Ellis returned to the laboratory from Rutherford, he spread the news of what he had seen—the Gospel of dijets. The word rippled through the corridors that supersymmetry was on the verge of being realized. Ellis had a Cheshire cat–like contentedness. "The experimental facts are coming out," he told me, "and they'll speak for themselves. I think that the best explanation of the facts so far is supersymmetry. And supersymmetry is the way of the future, maybe even the present. I think that finally people are just going to be convinced by experiment. They are not going to be convinced by some theorist shooting his mouth off."

Dimitri Nanopoulos was ecstatic: "If this is correct, this is going to be spectacular. This is no joke. It's not any crappy new quark. It's a really new thing."

With the theory itself looking good, the theorists took to fighting more vigorously over the nuances. Whose particular version more accurately mirrored the data? Even de Rújula was involved now. Four months earlier he had been comparing supersymmetry to alchemy; now he was lecturing on the subject and his own work therein. Nanopoulos cited the transformation as proof of the theory's validity. "What more success do we need now?" he said to me. "We just play it cool."

Three versions of the theory, propagated by three groups of theorists, were floating around the physics journals by this time. They differed mainly in their speculations on the masses of the various undiscovered SUSY particles, or sparticles, as they were called. The mathematics of supersymmetry provided a rough range, but beyond that the theorists could pretty much pick different masses, play with their scenarios with a technique they called event generating, and see what came out.

The game was to find a particular set of masses for the sparticles that would give you as closely as possible the events that UA1 had reported. Ellis and Nanopoulos had decided that their best bet was with squarks weighing 40 GeV and gluinos weighing 40 GeV. Ellis said the choice was based simply on gut feeling. De Rújula had been working with the Spanish contingent—the ones who had started the ball rolling by predicting monojets—and was playing with gluinos that weighed a few GeV and squarks that weighed around 150. The Wisconsin contingent also liked light gluinos and heavy squarks that weighed around 100 GeV, but they had other assumptions that differed from those of the Spaniards. Depending on which combination of masses they chose, their event generators would predict that there should be a certain number of dijets for every monojet in the UA1 data, or that the actual jets in these events should be within a certain range of width—either narrow or wide.

The theorists were then left to fight over whether the handful of events from UA1—the reality of which the experimentalists themselves could not even agree on—seemed to have wide jets and more monojets than dijets, or narrow jets and fewer monojets than dijets, or something in between. It was like watching the proverbial blind men argue over a photograph of an elephant.

When Ellis had lectured in November, he had presented his particular model, which he called a scenario, in the best possible light, and had been heckled by de Rújula throughout for doing so. Ellis's title was "How to Persuade Susy to Reveal Herself." The byplay that ensued between him and de Rújula was so vitriolic that much of the audience—particularly the younger physicists—became as interested in it as in the physics of Ellis's talk.

When Ellis suggested that he would ignore particular events that could not be explained by his model, de Rújula interrupted: "So when the events do not agree with supersymmetry, they are wrong. It's not the other way around?"

"Exactly," Ellis countered. "All of those who remember 1974 [he was referring to the high-y anomaly, on which de Rújula spent an unproductive year with Glashow, and which Ellis ignored as most likely incorrect] will know that one should not look at the data without a certain amount of intelligence."

"That's what I was saying," de Rújula returned after a long pause, provoking laughter from the audience.

"Different theorists," Ellis said after another long pause, "have different intelligences."

Later, Ellis mentioned a variety of W decays that might explain the monojets as standard model, but he refained from showing them.

"All of these are compatible with the data?" de Rújula asked. "Yes."

"That's why you're not showing them?" de Rújula returned.

"I told you," Ellis countered, "I'm only going to push my shit, Alvaro."

Forty-five minutes into the lecture, Ellis seemed to lose patience with de Rújula's heckling. He had stayed even, but he was tiring while de Rújula was still fresh. He simply scowled in response to the Spaniards further interruptions.

In January de Rújula presented his own lecture, entitled "The Madrid Model Confronts the Collider Monsters." The CERN physicists looked forward to a return engagement, but Ellis was lecturing in Argentina. With a cigarette dangling from his lips, his hands in his pockets, de Rújula informed the audience that the title of his talk would not be "Supersymmetry Is Found." His style was to alternate classical calculations with biting sarcasm. He first attacked the techniques of his theorist colleagues. "For the sake of discussion," he said in his accented English, "if one is in the game of attempting to discuss whether SUSY has anything to do with data, one has to place oneself in the somewhat unpleasant situation where one refers to some events and not to others. Because SUSY is not very nice to explain any of the others. So one has to hope that of all this rubbish, everything disappears in the future but these events . . . I have no justification for doing this, other than tradition."

He then attacked the jargon of his theorist colleagues. "The fact is the following," he said. "When I was younger, and most of you were also younger, the examples of theoretical concoctions were called models. That is a fact of the past. They are now called scenarios. This is not by chance. This is a Freudian reflection of the move from theoretical toward theatrical physics."

As he spoke, he flashed his transparencies on the screen. They were written by hand in yellow, blue, red, brown, green, purple, and black characters. "It's easier to keep awake," de Rújula told me later, "when you're looking at colorful things than when you're looking at black and white." Meanwhile, he ran down the other models. He explained how UA1 had set limits on the gluino masses at greater than 40 GeV, using the data and Ellis's model. And how his colleagues had used this to assert the correctness of their model over the others. He then calculated how they had erred in their analysis, and came to the conclusion that the 40 GeV boundary was premature. "The correct statement," he said, with a deadpan expression, "would be that the mass of the gluino is either more than 20 GeV or less than 20 GeV."

This, therefore, left a rather large window for other models, including the Madrid model of the Spaniards, and the Wisconsin model, the authors of which de Rújula referred to as the Wisconsin yo-yos. He ended by claiming that the Madrid model was "slightly less bad than the other scenarios," and then adding, "but this is a question on which one can have one's own opinion." (When de Rújula presented roughly the same lecture in Boston a month later, Glashow, his close friend, told him that "the degree of viciousness was beyond what we are normally accustomed to in seminars.")

Even the most pessimistic and cynical of the theorists were now beginning to believe that the UA1 data, which they had yet to see in its entirety and for which the background analysis had yet to be done, confirmed the predictions of supersymmetry. Even Glashow had become less vituperative than usual on the subject. He refused to give in. ("Of course they can't discount supersymmetry," he still insisted. "With so many particles, supersymmetry can say anything you want it to say.") But he was looking forward to the February meeting in St. Vincent when the data was expected to be released, and he was ready, if need be, to make his speech of conciliation.

On January 17, at about ten-fifteen in the evening, Felicitas Pauss finished scanning the last ten data tapes of the 1984 season. She was alone in the Megatek room, Rohlf, Mohammadi, and Revol being in the States.

The final Z was found that night; the fifteenth candidate. The last W was in there, too. They had 114 candidates for the decay of the W into an electron and a neutrino. They had found no more mono-jets or dijets of interest in the last third of the data. It was a little worrisome.

Out of the billions of collisions that had occurred at the heart of UA1, they had taken 4.4 million onto computer tape, and had chosen 550,000 of these to be processed quickly as potentially interesting. From these, the computer had sifted out 16,500 for scanning by eye on the Megatek, and from these, 5,300 events had missing energy that could mean new physics. After the first round of scanning, they had only 178 events that were not obviously junk created by cosmic rays, beam haloes, machine malfunctions, or particles slipping through cracks in the detector. Of these, they had eight new mono-jets and two dijets with missing energy greater than 40 GeV, the cut-off they had chosen in the 1983 paper to denote potentially new physics.

Those were the raw numbers, nothing more. They would have to be checked and rechecked. But the message had already gone out that the new physics had been confirmed. The UA1 collaboration would now have to calculate how many of these events could be explained by background. That was the hardest task. Taking the data was almost easy by comparison.

14.

FEBRUARY, 1985:
BUMPS IN THE NIGHT

"I'm getting a little tired of the UA1 collaboration and their complaints about Carlo. If they don't like Carlo, they should kick him out. And if they don't kick him out, it's because they need him."
— Don Perkins, professor of physics, Oxford University

In mid-January, Rubbia decided that they would analyze the new data in Boston and not at CERN. He had a course to teach at Harvard in the spring semester, as did Rohlf. Revol had his first classes at MIT. Geer and Pauss would join these three in Boston, and, with graduate students to help, they would carefully rescan all the events. They would work out the possible backgrounds and, they hoped, finish at least the preliminary analysis by the last week in February for the Proton-Antiproton Workshop in St. Vincent.

The workshop was critical: The physics community expected Rubbia to present the analysis of the monojets and dijets there. St. Vincent was no longer a workshop, as it had started out. It was now perceived as the definitive conference of the year. It was the deadline.

The first weeks in Boston were discouraging.

I had flown there to follow the analysis and found Rohlf and Revol on a Saturday evening in the Megatek room at the high-energy physics department at Harvard. The building is at the northern tip of the campus, where Harvard and residential Cambridge merge, a collage of old wooden homes and red brick buildings. Rohlf was working on his class notes. Revol was running tapes for the analysis. They were waiting to get a piece of computer code written at CERN that would fix a bug in their analysis program. In the meantime, they had been working on their classes.

Now they told me there was no rush to do the analysis. It was more important to do it well. Geer hadn't come. Pauss had been sick and hadn't come. Rohlf looked a bit saddened by it all. Maybe it was the Boston winter. Maybe it was the slow Saturday nights. They both had more friends at CERN.

Rohlf was teaching an undergraduate course on quantum mechanics that met three times a week and took up all his time. During previous semesters he had tried to teach while commuting regularly back to CERN. As a result he had garnered a reputation as one of the more poorly prepared lecturers on campus. ("Students do not doubt his expertise or his concern for them," wrote the *Harvard Crimson's Confidential Guide,* a handbook to courses at Harvard, "but they felt he presents the material poorly.") Rubbia had been able to pull this off, but only with help. A tenured professor on his way to a Nobel Prize could afford to skip classes, or ask his assistants to fill in. Rohlf had no such fallback. And he knew that regardless of his skills as a physicist, he would have trouble getting tenure if he did not improve his class work. "I'm already bored," he told me, while writing up lecture notes on a word processor. But he kept at it.

Revol was teaching six hours of classes—eighty students in total —and he was worried how he was going to handle even that. He was already exhausted, and was hoping he could recruit grad students in time to bail him out. With Rohlf, he was just getting started on the analysis. Now, he said, they had no deadlines. It could be months.

By the middle of the month, it was obvious that the analysis was running far behind schedule. But there was much more to it than the burden of teaching shared by Rubbia's disciples: The political situation at UA1 had blown up.

On February 4, UA1 held its first physics meeting since the Tau meeting two months earlier.* The aim of physics meetings, as opposed to the six o'clock meetings, working group meetings, or executive committee meetings, was simply to bring the entire collaboration together to hash out the important physics topics. Collaboration members would fly in from England, Italy, Austria, France or wherever they happened to be. They would discuss physics, and then fly home.

The meeting room quickly filled to overflowing for the two-thirty meeting. Rubbia uncharacteristically arrived precisely on time. He was in a gracious mood, smiling, accepting the commotion with unusual good humor. The UA1 members joked and babbled and

---

*I was still in Cambridge. This account of the meeting was assembled from interviews with the physicists involved.

waited for the physics, which as it turned out they would never get to hear.

When the room had settled, Rubbia explained that he had reviewed the topics to be covered and that the list was too long. It could only be covered, he said, if they stayed until midnight, and even if then, they would never do the physics justice. He suggested that they save half the topics for a second meeting at a later date. Then he told them he would make an impromptu agenda, because the agenda was properly the responsibility of his co-spokesman, Antoine Leveque, and he did not know what Leveque had in mind. Leveque, Rubbia told them, had been called away because his mother-in-law had died. She's dead, Rubbia said, deceased. And he reiterated it four or five times.

Acting the parliamentarian, Rubbia asked for motions from the floor: Should they try to discuss all the topics, or should they put half off for another day?

Many of the physicists began to wonder what was brewing. They had seen Rubbia use such excuses before. It was generally understood that Rubbia made every important decision pertaining to the experiment, and never went into a physics meeting unprepared.

Daniel Denegri suspected what was going on. That morning, Rubbia had argued with a young Saclay physicist, Elizabeth Locci, about whether she would present her physics work in the meeting, or simply her work on the calibration of the calorimeters. Several UA1 physicists had seen Locci come out of his office, apparently infuriated, her graduate student in tears. Whatever had happened, Rubbia had made his point.

It was the rule of the collaboration. You build the apparatus, you maintain it, you recalibrate it with your. The calorimeters were French, so the French had to do the calibration. It was as critical as it was difficult and painstaking. Without accurate calibration, for instance, the energy of a W particle that the detector recorded at 80 GeV might actually be 70 or 90 or anywhere in between. More importantly, a monojet that appeared to have too much missing energy to be the product of any phenomenon of the standard model might simply be mismeasured. There was no way to tell until the machine was carefully and thoroughly calibrated. Rubbia needed the French calibration work, and that's what he wanted to hear reported.

Denegri knew that Rubbia's disciples had taken the missing-energy analysis back to Boston. If previous history was anything to go by, this would be the analysis that would be discussed in conferences and written into papers. So he assumed that Rubbia had resolved to review only half the topics in the meeting so as to push the Saclay physics work back to the next meeting, which was as yet unscheduled.

S. Denegri told Rubbia that the French physicists had the right to present physics data, not just calibration data. And Rubbia accused him of trying to turn a physics meeting into a political arena. Later, many of the UA1 physicists would agree with this. Denegri, they said, was out of line. But at the same time, those who knew how deeply the clashes between Rubbia and the Paris contingent had cut, knew that Denegri had to force the issue if he were ever to get a fair hearing. Denegri no longer had Rubbia's ear. If he tried once again to talk to him quietly, he would fail, or so he believed. And he was getting angry.

"It's not fair," Denegri said to Rubbia, his voice rising, "that Felicitas Pauss, Steve Geer, and Jim Rohlf are the only people who are allowed to do physics in UA1."

Rubbia whirled around to the blackboard, and on one side scrawled in large letters the names of Felicitas Pauss, Jim Rohlf, and Steve Geer. On the other side, he wrote "Fair" and circled it. Then he turned toward Denegri and, jabbing his finger at Denegri's face, said: *"You* are accusing me of being unfair. That's a serious accusation."

The show was on. Rubbia asked again how Denegri could claim that Rubbia was attempting to prevent them from speaking, when the agenda had yet to be set. Denegri replied that Rubbia had told Locci that morning that she would be allowed to present only calibration data. Rubbia pivoted, searching for Locci, and when he found her, he asked her to say whether this was true. She calmly replied that he had said to her, no physics . . .

Rubbia interrupted. *"No,"* he said, refusing to submit, "don't tell me what I told you."

Locci, under attack, defended herself. She was confused, and maybe a bit scared, but she refused to back down. What she did next pitched the meeting against Rubbia. He began to lose control. "Now you have insulted me," she told him, "so I have the right to say what I want to say." In the years that she had worked in UA1, she said, Rubbia had kept her from doing physics that she was capable of doing. She was no longer willing to do only what he told her to do. She wanted to present the physics, but she had been kept from doing so because Rubbia refused to let her discuss it in group meetings, saying she had not first briefed him in private. Yet he never seemed to find time to discuss physics with her in private. "That's all I wanted to say," she concluded, "and if you have any answer to that, go ahead."

Rubbia asked if others in the group felt that they were prevented from doing physics. He turned to a Collège de France physicist standing in the back, who had walked in just a few minutes earlier. Rubbia asked him if he agreed with Locci, and the Frenchman said yes. Rubbia was flabbergasted. "How could you agree?" he protested. "You don't even know what's being discussed."

"Basically eighty percent of the audience was on our side," Denegri said later. "This had never happened before. Usually when Rubbia accuses somebody, nobody is on your side. In this case, it was an absolutely unique situation."

Rubbia had lost control. This is ridiculous, he told them, attempting one last rally. I will not subject my group to such a display. He threatened to postpone the meeting until such time as the situation could be clarified. Pointing again at Denegri, Rubbia said that the man had made a serious charge against him and against the three people whose names he had written on the board. He wrote his own name on the board under the list of three people. "Rubbia." We have been insulted as a group, he said, and until these sorts of problems can be solved as a group, we can't discuss physics on any realistic basis.

He walked over to Felicitas Pauss, who sat with her elbows on her knees and her head in her hands. He asked her if she thought she had been insulted. Pauss smiled sadly, shook her head, and said nothing. Rubbia walked back to the board. A serious charge, he repeated. It would have to be discussed at the level of an executive committee meeting, and until then they could not talk about physics.

The meeting was over. Giorgi Salvini, the elder statesman from Italy, tried to salvage it and failed. Horst Wahl, the leader of the Austrian contingent, tried to salvage it, was insulted by Rubbia, and left. Rubbia had decreed that there would be no physics. But his was not the victory.

The following day at CERN, Rubbia held a meeting of the physicists working on the missing-energy analysis. Rohlf and Revol were still in Boston, trapped by their classes until the end of the semester, and so Rubbia assigned Alan Honma to take charge of the group.

Honma was the kind of physicist that others went to when they had a problem. But more than anything, Rubbia had assigned him to take over the analysis because he was not one of the "terrible three," as Rubbia had already in jest begun to call Pauss, Geer, and Rohlf. (The "Gang of Four" when he included himself.)

Honma was a young American who had done his undergraduate work at the University of Michigan and had then gone on to SLAC for his doctorate. At the time he started with UA1, he also took a research position with Queen Mary College in London. Like many physicists with temporary research appointments, he had yet to visit the campus. Honma was tall, dark, with long thick black hair. He was extremely self-contained and competent. After three years at the experiment he was virtually running the trigger system. "Alan sort of does all the work," Eric Eisenhandler had told me. "It's hard working with him, because he can do most things faster than most people. He doesn't tell you what he's doing. He just does it. You keep having to pump him. If he starts to tell you what he's doing, then you know he's worried and in trouble." Even Rubbia was impressed, and had learned that Honma should be both respected and listened to when he did actually say something. During the run it seemed that Honma could be found twenty-four hours a day on the triggers in the control room. Then, as the run wound down, he had moved to the missing-energy analysis.

After the confrontation at the physics meeting there was no longer any question of the monojet analysis being done in Boston. Now it was back at CERN. Pauss and Mohammadi were still working hard, and they had recruited another young Englishman, Richard Batley, who had been working on the triggers with Honma. And Rubbia had finally tried to integrate the Paris physicists into the analysis in some coherent way, but only on condition that all the work be done together, without anyone running off on their own and then coming back the day before St. Vincent with individual results. Rubbia made it very clear that if anyone wanted to work on the monojets, they had to stay in town, and they had to work at his pace.

The aftershocks of the confrontation reverberated around UA1 for some time. By the middle of February the political situation had reached critical mass. On February 13, Rubbia sent a Telex to the members of his executive committee. They would have an emergency meeting on the 20th. They must all be present. Substitutes would not be accepted. The physicists referred to it as the "Future of UA1 Meeting."

A few days before the meeting, Cline came back from the States and confided to Mohammadi and a Wisconsin post-doc that Rubbia had told him he was considering taking a one-year sabbatical from UA1. His reason, according to Cline, was that the collaboration didn't realize his true worth; he was tired of the fighting, the politics, and the day-to-day grind. Cline discounted Rubbia's claims that he needed a rest, and instead suggested that he wanted the year off to get the proton-decay experiment cranked up in Zichichi's Gran Sasso lab in Italy. Earlier in the month the Italian government had suggested that it was ready to double the budget for physics research in Italy. A good deal of the boost for that money had come from Rubbia's Nobel Prize and his near cult-hero status in Italy. (In February he was featured on the cover of *L'Uomo Vogue,* a swank Italian fashion magazine.) Rubbia and Cline took this influx of cash to mean that the Italian government would back their proton-decay dreams if they settled in Gran Sasso. "It seems like if the money is available," Cline said, "then you should not piss your time away."

It was a grandiose project. "Whether it could be made to work," Cline told me, "I don't know, because it's a hell of an ambitious project. But it has the right smell to it. It looks a little bit like the collider project. That seemed impossible too."

The rumor that Rubbia was contemplating a sabbatical spread through the lab, accompanied by the usual speculations. The day before the Future of UA1 Meeting began, the physicists heard that Rubbia had told the director general that he wanted to resign as spokesman of UA1, but would stay on as leader of the CERN physicists working in the collaboration. Like Cline, the other physicists gave little credence to the idea that Rubbia might simply be burnt out. But, unlike Cline, they believed that in fact he was reacting to the latest news from the management. The CERN Council had just announced that it had reelected Herwig Schopper as director general for three more years, an unprecedented extension of his term. Concurrently, they announced the establishment of a "Working Group on the Scientific and Technological Future of CERN." Its mission would be "to explore various options for the long-term future of CERN," and its chairman would be Carlo Rubbia.

The council had actually voted on the director general position in December, but at the time the Italian delegation had resolutely insisted on voting for Rubbia. CERN had never had an Italian DG, and the Italians felt that with the Nobel Prize the time had come. But the council preferred that the decision be unanimous, and postponed the official announcement until a compromise could be found to appease the Italians.

The council would not make Rubbia director general, even though, as one former lab director put it, "Schopper is a good politician but he is an extremely bad psychologist. And nobody likes him." The council members must have feared, as did the physicists and technicians, that should Rubbia be made DG immediately, he would cancel LEP, the huge electron-positron machine in the works at a cost of half a billion dollars, and instead arrange to install the LHC, a second-generation proton-antiproton collider.*

When the council reelected Schopper, they ensured that at least the first stage of LEP would be completed. Offering Rubbia the chairmanship of the working group on the future of CERN was the equivalent of making him director general designate. It would placate Rubbia and the Italians. It also represented a green light for Rubbia to put into motion the political machinations that would be needed to bring about the LHC. (Cline, for one, figured it meant the beginning of another ignominious defeat for the Americans in high-energy physics.) All Rubbia had to do was find the time to dig up several hundred million dollars in the next five years and create a new accelerator. It was quite conceivable that he might want to quit UA1 to do it.

The day of the Future of UA1 Meeting, Rubbia seemed nonchalant. He went to lunch with Pauss and came back apparently in a good

---

*The politics behind this story spans two decades of physics. Grossly simplified, it goes like this: In 1981, before the collider had strutted its impressive stuff, CERN set plans in motion to build LEP. Shortly thereafter, Burt Richter at SLAC started out on his own electron-positron machine, revamping his old linear accelerator into a machine that would be as powerful as that of the European competition.

When Rubbia's collider came through, the Americans decided they wanted to build the Superconducting Super Collider, or SSC. Concurrently, Rubbia and the Europeans decided that they could build a machine just like it, known as the Large Hadron Collider, or LHC. It would be only half as powerful as the SSC, but might be built three to five years sooner—or at least so Rubbia opined. The Americans talked about 1995 for their SSC. Rubbia thought he could put an LHC in the LEP tunnel by 1991, for a quarter of a billion dollars, although the Americans insisted that an LHC would cost Rubbia $3 to $4 billion, and that he wouldn't get it in until 1995.

When it became clear that Richter's machine would be finished two years before LEP, LEP appeared to be redundant. But LEP was scheduled to be built in two stages, the second of which would be twice as powerful as the SLAC machine. LEP was also a workhorse that would provide jobs as much as physics. It would have four experiments on the single ring (none of which would be Rubbia's) each with several hundred physicists. Rubbia had said that he thought CERN should reconsider and abort, at least, the second stage of LEP in favor of his Large Hadron Collider. On the subject of LEP, Rubbia had been known to say that he "didn't want to play the trumpet in the Salvation Army."

mood. He stopped in his office long enough to fire his secretary yet again. She said it was maybe the millionth time he had done so. She was not worried: He had always reconsidered in the past. Then he walked into the executive meeting. When it broke up a couple of hours later, no one involved would say what had happened. But everyone looked a little grim.

The only news that trickled down to the lower echelons was that a young post-doc who usually took the minutes was told not to, and that the meeting was restrained. But the mere fact that it was kept a mystery meant to many that something drastic had happened. They assumed that eventually Rubbia would make a statement.

The next day, Leveque had coffee after lunch with Pauss, Mohammadi, and Mike Levi, a young American physicist. Levi was small, with short blond hair and eyes that always seemed to be laughing at some private joke. He had an irreverent style and little or no respect for bureaucratic convention. Levi turned to Leveque and said bluntly, "So what happened at the exec meeting yesterday?"

Leveque answered that Rubbia had informed them he would resign as spokesman at the end of the year. The others had told Rubbia that this was unfortunate, and asked him to reconsider. They told him that if he left immediately, it would be "catastrophic" for the UA1 collaboration and even for CERN.

"Carlo does care what happens," Leveque insisted. "He does want to see UA1 do good physics. He doesn't want to see it fail." But the others couldn't help wondering if he was actually about to abandon UA1 in the same way he had abandoned his previous experiments.* They had a strong feeling of déjà vu.

The previous November, CERN had approved an upgrade of the UA1 experiment, which included the new uranium calorimeters, at a cost of 24 million Swiss francs (at the time roughly $9 million). Many of the UA1 physicists were scared already that the upgrade would be a debacle. They felt that the UA1 proposal had been a bit halfbaked—it had been done quickly—but that the CERN committee that reviewed it would force them to tighten it up. Rubbia was

---

*The next day I talked to Don Perkins, the Oxford physicist who had known Rubbia for twenty-five years. I asked him if he thought Rubbia would quit. "When Carlo leaves UA1," he told me, "I'll eat my hat publicly. He might try to renounce leadership, but he'll still keep his nose in and he'll still be beating the rest of the collaboration over the head to get this done, get that done."

known to advertise the most optimistic numbers for his plans, but in the past someone had usually been around to exercise moderation. This time, however, the committee, probably betting on Rubbia's recent track record, had given the entire proposal the green light. "Some people took the view," one British physicist said, "that we put in some things as a bargaining ploy so that when the committee then queried everything, we'd say, okay we'll simplify it and cut them out. But the committee didn't ask us anything. They just waved it through, and we're stuck with it."

A French physicist had told me that several members of the committee confided that they had approved the UA1 upgrade, and given Rubbia all that money, with the idea that if he actually pulled it off, then he deserved every penny of it, and if he failed, it was a way of letting him commit political suicide.

Unfortunately, if the upgrade failed, the rest of the UA1 physicists would go down with Rubbia. But when they tried to discuss the upgrade with Rubbia, he was never around enough to pay it any attention. And they could do nothing that was not certified by him. "We have serious problems," the Englishman said. "We want to talk to Carlo about it, and he's always busy. And his response is always either a joke or a short, sharp insult, and it is not a serious discussion. And I'm quite worried. I think the upgrade could go off the rails."

The microvertex detector was the perfect example of Rubbia's over-optimism. Rubbia had conceived of it in 1983, to help nail down the top quark and maybe pick up the Higgs boson into the bargain. The device would fit inside the central detector and be the most precise high-energy physics apparatus in the world. As with the central detector, SLAC physicists had built one first, but they had taken three years with it. Rubbia wanted his to be more precise and to be built in a year. First he asked Mario Calvetti—who had helped build the CD and had bailed Rubbia out on the electronics—to give him the names of people with whom he could build the microvertex in a year. Calvetti chose the best chamber builders in the experiment. Then, as Calvetti began the design, Rubbia gave the job to Anne Kernan's Riverside group, because they had the money, and because he wanted to demonstrate to the DOE that U.S. groups played a major role in UA1. And, so Calvetti and Sadoulet believe, because he wanted to take credit for the design.

Kernan took the job because if the microvertex could be done, it would be prime physics. Until then, the Riverside group had been only obliquely connected with the analysis. But Kernan was a bubble-chamber physicist with no hardware skill, and she had only a few technicians, a couple of young inexperienced post-docs, and some CERN support to do the job that Calvetti had wanted to do with the best chamber people around. Calvetti, in the meantime, quit in disgust and told me he would never believe Rubbia again. "In his mind," Calvetti told me, "Carlo thinks that he does not need brains. He needs only hands."

Rubbia gave Kernan the design and nine months to build it. It would have to be installed and running for the 1984 run. When they heard about this assignment, several of the technicians immediately suggested it was impossible. But pessimism wasn't allowed, since it was Rubbia who had said it could be done. Kernan later confided that she thought the odds of pulling it off were at best one in a hundred, but that if the young people believed it could be done, they would work all that much harder. The DOE gave Kernan half a million dollars for the chamber.

They then worked for nine months, seven days a week. And they failed. The microvertex had problems—wires broke, sparks shot out. Much of the fault lay in Rubbia's original design and his belief that the most sophisticated apparatus of its kind in physics could be built by a group of kids who had no experience.

When they couldn't get the microvertex in as promised, Rubbia blamed it on the fact that a special beryllium component had been held up for three weeks in U.S. Customs due to President Reagan's decree against exporting high technology to countries with eastern bloc connections. This was what Rubbia told me in October of 1984. One of the younger technicians on the microvertex called this "bull shit." Another Riverside physicist laughed when I mentioned it. "That's our party line to the DOE," the physicist said. "We have to stick to that story now."

The group then received more DOE money for a second microvertex. This was started in the fall of 1984, and the post-docs and technicians put in another nine months of twenty-hour days. This version too had problems, but eventually it would be installed for the run that began in September 1985. It would run for ten days before developing what one technician called "extremely frightening things" that forced its removal from the detector.*

Now Rubbia had told his executive committee that he was going

*The collider was shut down for a week while UA1 was removed. UA1 was shut down for nearly a month while the microvertex was removed.

to quit at the end of the year. In fact, he had already put the plan in motion. Those physicists working on the upgrade began to see their projects going the way of the microvertex. They felt that they were witnessing the beginning of the end.

The disciples didn't seem to care, however, for by then the physics analysis had taken a bizarre and promising turn of its own. It had been obvious that the results from the missing-energy analysis would simply not be ready for St. Vincent. Honma's working group had too few bodies and too much politics to worry about. Pauss and Honma had been rescanning events, verifying that they had found all the possible monojet and dijet events, which was a long and tedious job. But then the Bump arrived.

It was exactly that: a bump in what should have been a smooth line in the curve of the mass distribution of the W's. Because of the vagaries of the experiment, the masses of the W's were never measured exactly at 82 GeV—the mass as predicted by theory. When the physicists plotted the curve of all the masses, if everything worked, they would have a smooth curve that peaked at the predicted 82 GeV. But when they analyzed the new data, they found a second peak in the curve, a bump in the distribution. The Bump. Something seemed to have generated more events at one particular mass than at the others, and it was not the predicted mass. That something could be background, or it could be a new particle. It had been in the data the previous year, but they had ignored it. "We thought it was garbage," Rubbia told me, "just statistics. And we hoped like hell it would go away. This year, with more statistics, the thing's huge. What are we supposed to do?"

Shortly after the Bump made its appearance, Rubbia went through his files and unearthed a paper that dated back to July, 1983. It had been written by three Italian theorists, and it predicted that if the supersymmetric partners of electrons—called selectrons— were made in the collider, they might be found in just such a bump. The next day Rubbia convened a meeting to discuss the possibility of looking for supersymmetry. He assigned a young Finnish physicist to begin simulating supersymmetric decays with a Monte Carlo. He wanted to know what to expect. And he wanted to know before St. Vincent.

If the Bump were real, and if it could be attributed to some new SUSY particle, then it would be exactly the confirmation UA1 needed for the missing-energy events. Provided, of course, that the missing-energy events were still real. If they could find two different supersymmetric particles in two entirely different analyses, the world would almost have to believe them. Rubbia was so excited about this that when he and Cline went to Washington to talk to the government about their proton-decay detector, Rubbia showed it to Cline, who had not yet seen it, and described it as a "bombshell."

At UA1, they were working frantically to figure out the background on the Bump before the St. Vincent meeting, which was now five days away.

On February 19, Rubbia attended a CERN Scientific Policy Committee meeting at which UA2 showed their plot of the W mass distribution to the CERN management. This plot was identical to that of UA1, but UA2 had begun their plot at the point where the Bump started on UA1's plot. UA2 might have had such heavy background in that region that they just hadn't bothered to include that region. Or they might have discovered the Bump and decided to keep it secret until they could blow the whole thing open at St. Vincent. One way or the other, Rubbia wanted his Bump kept quiet. The last thing he needed was for the word to leak out so that UA2 could rush off and look for the Bump themselves. "I consider it absolutely mandatory that nothing said gets out of this room," Rubbia told his physicists at one of the meetings. "I will deliberately try to follow any leak that comes back to me and go back to the origin: like the CIA or the KGB. All right?"

Steve Geer was leading the UA1 analysis, since it fell under the category of W physics that he had been concentrating on for the past year. He had help from Mohammadi, Levi, and Rajendran Raja, a thirty-six-year-old Cambridge-educated physicist on a one-year sabbatical from Fermilab. A group from Saclay were rushing the calibration work, which was crucial for the Bump.

Geer was spending all his time rescanning the W events. To reliably measure the mass of the W, they had to use only "gold-plated" events. Even minor problems with the events would throw off the measurement of the mass. During the run, the W's had been taken with two different triggers. One was the electron trigger, in which the detector saved the event if it spotted a good electron in the collision. They had found two thirds of their W's this way, and when Geer rescanned these events, almost all came through as gold-plated. The remaining third of the W candidates had been taken by the missing-energy trigger, and this was worrisome. The missing-energy trigger cared only whether there had been some imbalance in energy in the collision. That trigger would find W's that had been missed in the electron trigger, but it would also scoop up a wide variety of junk, including the very bizarre junk that imitated W's.

After rescanning the W's from the electron trigger, Geer realized that most of the W candidates that made up the Bump had been caught with the missing-energy trigger. This could mean that the Bump was mostly junk, simply fake W's with fake electrons.

On February 20, Geer reported that these events had to be rescanned with "extreme urgency. It seems we rescanned the wrong events first." He figured they had two more days of scanning, which they could comfortably complete within the three days before the St. Vincent meeting.

On the 21st, Geer reported that in rescanning the W's, he had thrown out more of the events that made up the Bump. They were fake W's, and as a consequence the Bump had receded a little more. The work went on.

They were waiting for Mohammadi's Monte Carlo. He had been working desperately to come up with a simulated mass plot for the W that incorporated his tau decays. When the W decayed into a tau and a neutrino, the tau could in turn decay into an electron and a second neutrino. In this case the physicists would see the electron and the missing energy as though it were from the original W, but in fact it wouldn't be. A large number of these W to tau decays could conceivably account for that second peak in the mass plot. But no one knew for sure because they simply hadn't bothered with it the year before.

Mohammadi's preliminary Monte Carlo seemed to indicate that background could account for only a handful of events, and the peak, although it was shrinking, seemed to be inflated with well over a handful. They needed a final sample of gold-plated W's only and a final Monte Carlo.

Rubbia pushed Mohammadi to finish the Monte Carlo. If Mohammadi ran off the same number of W's as they had used in the mass plot, they could take the curve that came out of his Monte Carlo and simply lay it over the curve from the data. Just by eyeballing it, they'd have a good idea of whether they had a bump that the standard model couldn't account for.

At the meeting, Rubbia asked Mohammadi if he could do it.

"Yes," Mohammadi replied.

"By tomorrow?"

"No," Mohammadi said.

"Why not?" Rubbia asked.

"Because high statistics takes a lot of time."

"Do it anyway," said Rubbia.

With two more working days before St. Vincent, Rubbia summed up the status of the Bump. "Now, what are we going to do with that Bump?" he asked his physicists. "Is it garbage? It is an important and difficult decision which I do not believe we can make a priori. Nobody with a machine gun will force us to come out with this, unless by chance the other guys come out with it and force a shoot-out."

On February 22, Rubbia met with the physicists still slugging away at the monojets and took a vote on whether to discuss that subject at St. Vincent. Rubbia looked grim when he entered the conference room, self-absorbed.

"It has been decided by the executive committee," he told the roomful of physicists, "that we should discuss what should be presented at the meeting in St. Vincent. We have made a deliberate decision to slow down the analysis in order to make a deeper analysis and to spread the work to outside groups. This is healthy for the group, but slows things down. As far as 1984 data is concerned, I doubt that we will have very much to say."

He walked slowly around the room with his hands in his back pockets. He was calmer than usual.

"If I understand Alan [Honma] and Felicitas [Pauss] correctly, we have a sample of missing-energy data which is very rich, which will allow us to say more or less what occurred in this range last year. We can say that if anything was done wrong last year, it will be wrong this year. We can say that the signal we have is a significant signal. What we do not have at the moment is a novel new calculation on the background. There are two possibilities. We can use last year's background estimate and patch it up and correct things, and say we have a signal and the background is small. Say we still have work to be done and give a status report and say nothing more. The second possibility is to say nothing.

"We have to make a decision," he continued. "I propose that I shut up for a few minutes and let you people tell me what you think." Rubbia leaned on the table and banged his fist down a few times to make his point. Then he looked around inviting comment. Everyone was quiet.

"Nobody wants to say anything? Then let's start with the ladies first." He tapped Felicitas Pauss on the shoulder.

"I've thought about it," she said softly. "We have really nice data to look at. For me the only thing we can show is events to look at. I have a tendency to stay quiet."

"You will be interpreted in a negative way," Rubbia said pushing her a little. "People will ask questions."

Pauss said nothing.

"So your attitude is to shut up." And then to Aurore Savoy-Navarro, who was sitting next to her: "What about you?"

"I say the exact opposite."

"Thank you for being here," Rubbia said, without sarcasm.

"I would present a status report of what we have in missing-energy data," she continued. "Because, if nothing else, it's already been presented in Aspen and elsewhere."

"We're not committed to what has happened in the past," Rubbia said. "If that was a mistake, we could say it was a mistake, and not say anything."

Honma agreed with Pauss, saying that a status report would lead to too many questions.

The next physicist voted for a status report. And the next. Rubbia was speeding up, getting animated, pacing faster around the room. He was abrupt with anyone silly enough to ramble.

"Status report," said one.

"Keep quiet," said another.

"Keep quiet," said Denegri, finishing it off.

Rubbia grinned. It had been roughly even. "So we're in a pretty shitty mess, aren't we?" he said. "I cannot neglect the fact that people who are working on it have more weight than people who aren't. It's also clear that we cannot run science on a majority basis."

Later that day, he told the physicists that they would not talk about missing energy at St. Vincent, and he pulled his forces off the analysis and assigned them to the Bump. At 5:30 P.M. they met again on the Bump. After Geer's work, the Bump had shrunk further. But it was still there.

Mohammadi's Monte Carlo was still whirring through the computer. He had nothing to show. But the Finnish physicist who had been given the task of running SUSY Monte Carlo did have something. She showed a transparency of a simulated plot for a super-symmetric W decaying into an electron and a supersymmetric neu-trino. The plot showed two peaks, and at first glance they looked similar to the twin peaks in Geer's real W curve.

Rubbia walked up to the projector, took the plot in his hand, then rifled through the transparencies for Geer's plot. He placed them next to each other on the projector, and looked at them for a second. Then he put one on top of the other. They overlapped almost ex-actly. He appeared to shiver a little with excitement. "I don't know, folks," he said. "You make up your own minds. I'm getting nervous."

Mohammadi finished his Monte Carlo at four o'clock on Saturday morning. It was seven by the time he had his plot for the mass distribution. He left a copy each for Rubbia, Cline, Geer, and Levi, and went home.

At nine, Cline came in and noticed the plot. He overlaid the curve from Mohammadi's Monte Carlo onto the curve of Geer's actual W data. The two were identical. If Mohammadi's Monte Carlo was correct, it meant that the Bump could be easily explained by the W decaying to tau, and required no supersymmetric selec-trons. Cline walked over to Rubbia's office and tried to convince him that the Bump was not new physics. But Rubbia wouldn't listen to him. When Geer saw the original in Mohammadi's office, he took one look and said, "There goes the Bump."

At eleven-thirty the missing-energy crew met and Cline tried again with Rubbia. Rubbia told the group that there would be no talk in St. Vincent on the Bump, or the search for selectrons, as he called it. "But I suggest," he said, "we keep it in our pants. I propose we keep some of these transparencies with us just in case UA2 comes along with a big story on this. Everybody will turn to us and say, 'What do you have?' If UA2 has the story, I will take responsibility on myself to make a statement for UA1."

At this point, Cline deferentially tried to bring Mohammadi's Monte Carlo to his attention.

"Carlo," he said, "I don't mean to sound like I'm being very negative . . ."

"Dave, fuck it," Rubbia said, interrupting. "Forget it. I don't want to hear it. Okay?" He was waving his hands, chopping at the air.

"I just want to understand this curve," Cline said, holding up Mohammadi's transparency.

"Dave, excuse me. Don't give me physics now." Rubbia was

yelling, but he was not out of control yet. "Would you please stop forcing that in public. This is not the issue here. Enough. Okay? This is not for here. We cannot discuss this. We have to understand what we do."

And then he turned to the other physicists and changed the subject. "Listen, the time is short. Why don't we stop right now? Listen, how far are we with the calibration? What do you think?"

The rehearsal for the St. Vincent talks took place in mid-afternoon. There was a full house in the conference room. Rubbia arrived late. "Welcome to the madness," he said as he walked to the front of the room. Then he started in on one of the senior CERN physicists, Rudy Bock, who had presented him with the tentative schedule for St. Vincent. The organizing committee of the workshop, of which Bock was a member, had scheduled the UA2 W physics talk for Monday and the UA1 talk for the next day. These were the talks that would be about the Bump, if they talked about the Bump.

"I don't think this makes any sense," Rubbia said. "If this is not changed, I do not think we go. This program makes us look like the spare wheel on a car. We cannot do a thing like this. It's impossible. Either we get basic symmetry of UA1 and UA2 in these subjects or we boycott the program. I don't see any other choice."

"But, Carlo," Bock said, "UA2 hasn't even seen this yet. Let's discuss it after this meeting."

Then Rubbia began to set the UA1 schedule. Mike Levi would talk about W events that had jets. Rubbia chose Levi and not Geer for this, "in order to please Denegri, because Levi is not one of the terrible three." This was the talk that Levi called the damage control talk, the straight W and jet physics, without the Bump.

At midnight, twelve hours before they were to leave for St. Vincent, Mohammadi told me that Rubbia had still not acknowledged whether or not he agreed that the Monte Carlo had accounted for the Bump as background. "If Carlo really takes this as a final analysis," he said, "then nobody's going to worry about it. If not, then we're going to have to dig into it more and more." Twenty minutes later, the physicists were in Pauss's office. She had made espresso for the late-nighters. She had a box of truffles, a bottle of cream, and a bag of oranges. Pauss believed that if you had to work Saturday night, you might as well enjoy it. Moham-

madi was there, and two of the Saclay physicists, when Rubbia
walked in.

"Supersymmetry," he said, and as he said the word he dejectedly
whistled the air out of his cheeks and turned his right thumb up and
then swept it straight down. He had finally accepted the message
of Mohammadi's Monte Carlo. The short, volcanic life of the Bump
had ended.

15.

FEBRUARY 25–MARCH 1, 1985: THE FIVE DAYS OF ST. VINCENT

"I have to talk to you as an experimentalist, not as a theorist. I don't have a telephone line to God."
— Carlo Rubbia

No rumors of the political explosion at UA1 had leaked back to the States. The rumors that the monojet analysis might not be presented at St. Vincent had, however. Rohlf mentioned it to de Rújula while the Spaniard was visiting Boston, and de Rújula told his friend Sheldon Glashow. Glashow was incredulous, because the purpose of the meeting was to reveal just that new data.

Glashow came to an interesting conclusion: not only did he believe that supersymmetry was false, he also believed that the monojets of UA1 had been injudiciously claimed to be new physics. Rubbia had promised him results in January and he had yet to see them. If they would not be shown in St. Vincent either, it began to smell like the high-y anomaly to him. And once Glashow indicated that he believed the physics was false, the rumors began to spread.

St. Vincent is a small hillside town of three main streets, 100 miles down the Aosta Valley from Mont Blanc. It has an inordinate number of hotels with undersized rooms and occasional hot water, one luxury hotel, the Grand Hotel Billia, and a casino with two floors and enough roulette tables and slot machines to keep the entire population fully occupied. The Billia is an authentic prewar grand hotel. The conference took place in a modern addition in the rear of the building. Many of the young UA2 crowd showed up Sunday night in their ski outfits, having taken the opportunity to spend the weekend on the slopes. The UA1 physicists drove down Monday morning.

MONDAY

The Fifth Topical Workshop on Proton-Antiproton Collider Physics began at nine-thirty in the morning. The two hundred physicists present had come primarily to hear one thing. As Mario Greco, an Italian physicist, explained in his brief introduction: "The highlight of the 1984 Bern meeting was a few events which could not be accounted for in the standard model. These events generated enormous excitement in the physics community . . . and the new data taken this past year at the CERN collider will surely shed some light on the possibility of new physics at 100 GeV. . . ."

The schedule for Monday began with talks about new results from the UA1 and UA2 experiments. These held no surprises. Lou Mapelli of UA2 presented their results on W and Z physics. He showed the W mass distribution, revealing a small, barely noticeable bump in the same place that UA1's Bump had been. Mapelli ignored it. He concluded that their analysis of the W and Z gave results "as expected by the standard model."

Mike Levi followed with the classic W and Z and jets story later in the morning. His talk was a little rough, but it barely showed that he had not known he was going to give it until two days before, and had gotten less sleep in the last ten days than most people get in one good afternoon nap. In the question and answer period that followed the talk, Rubbia fielded most of the questions. Neither he nor Levi made any mention of any bump.

When the physicists broke for lunch, they learned that Rubbia had decided to give a short status report on the monojets the next day, even though he had been saying repeatedly that there would be no such talk.

TUESDAY

Rubbia spoke after the coffee break at 11:10 on Tuesday. "Now the fun begins," one of the theorists said to me. Rubbia began with a raw transparency, penned in the previous hour, that explained that he would discuss missing-energy events, searches for SUSY, and the top quark analysis. The transparency included a warning: He was about to show raw, preliminary data; they did not include background calculations or corrections for the efficiency of the UA1 detector. Theorists were not to use them for papers.

He proceeded to explain how the new missing-energy trigger worked, and gave the raw numbers for the 1984 run; fifty-four new monojets, thirty-eight dijets, and even ten events with three jets and one event with four—all with significant amounts of missing energy. All his numbers were as optimistic as possible. (For instance, Rubbia could have told the physicists that UA1 had fewer than a dozen monojets with missing energy greater than 40 GeV in a configuration that might represent new physics as it had been described in the 1984 monojet paper. Rather, he chose to say that they had fifty-four monojets and left unsaid the fact that the great majority were almost assuredly background or W to tau decays.) "We have a lot of material to work on," he said. "And this is why you don't have any reports here, because we really do have to think those things out in detail. Very complicated analysis. All that stuff has to be done."

Rubbia concluded with what he called his "Anomalies Scorecard" "just to show you where we stand in all our anomalies." He listed all the potential new physics from the 1983 run, stating which experiment—UA1 or UA2—had promulgated them. Next to all those effects that the UA2 physicists had pushed the previous year, Rubbia had scrawled "Not confirmed." Next to what they called the Z-gamma events, the odd decays of the Z with the extra photons that had started all the excitement about new physics more than a year and a half earlier, Rubbia had also written "Not confirmed." Next to missing-energy events, he had written "Confirmed."

The discussion period following was taken up with a dialogue between Rubbia and John Ellis on supersymmetric signals and how to find them. This led some of the audience to suspect that Rubbia was filibustering on supersymmetry in order to head off questions on the data. If this were so, he succeeded.

Whether Rubbia's talk helped UA1's credibility on the monojets was doubtful. First of all, Rubbia had flashed the famous missing-energy plot, which represented the tail of the distribution and showed the various monojets and dijets set down according to the amount of missing energy. The farther to the right on that plot the events fell, the greater the amount of missing energy, and the harder they would be to explain by the standard model. There were considerably more events on the plot when the 1984 data were added to those from 1983. Yet some of the physicists commented that even with Rubbia's enormous number of new events, they only seemed to fill in very nicely the hole between standard physics and the two most bizarre high-energy events of the last year. And that, they said, just made it appear more like the tail of the distribution of old physics, and nothing new.

The Anomalies Scorecard also did nothing for Rubbia's credibility. If the UA2 anomalies were not confirmed, as Rubbia had said, because that experiment had already calculated their background and admitted that it accounted for their signal, then how could Rubbia tell two hundred experienced physicists that the missing-energy events were confirmed when he had just finished telling them over and over again that UA1 had barely begun their background analysis?

This contradiction was taken with varying degrees of indignation. The older European physicists seemed not overly surprised. "Yeah, well . . . that's Carlo," was what DiLella said. On the other hand, one of the young UA1 physicists, who had been working on the missing-energy analysis, called it "disgusting." Then he told me that he could not understand how Rubbia had come to give the talk to begin with, and how or why he had all the transparencies with him, when they had been told there would be no talk.

The afternoon session on Tuesday was reserved for super-symmetry. Ellis began it, and was followed by Gordon Kane from the University of Michigan, who was in turn followed by de Rújula with the summary.

Ellis began by briefly summarizing his reasons for believing that supersymmetry perfectly solved the problems of modern physics. Then he covered the three SUSY scenarios that were floating around, giving the relevant predicted masses for gluinos and squarks. He went on to show quite a few colorful charts that were drawn carefully in accordance with various results from UA1 and UA2. These charts showed the number of "gold-plated" UA1 mono-jets from 1983 (at least five), the widths of the jets in those events ("really quite surprisingly thin"), and the number of expected dijets and even trijets ("You can ask yourself, can you get trijet events with missing energy from either squark, slepton, or gluino pair production?"). He also talked about what was known about the very beginning of the universe and the gravitational attraction of galaxies and other sundry quantities—astrophysicists could supposedly calculate how much photinos must weigh, if they exist, by inference from their observations of the universe. After admitting that "at this point we start getting into the business of over-interpreting the data," Ellis used all of this to illustrate the merits of his particular model and, as he put it, to "piss on the other guys."

This was followed by a colorful dialogue between Ellis and de Rújula, in which de Rújula accused Ellis of changing his predictions month by month for the masses of the sparticles in his scenario. Ellis responded that he did no such thing, but would not elaborate further. When de Rújula asked him to commit himself to one particular mass for the gluino, Ellis suggested they wait a few weeks to see what UA1 had to say. To which de Rújula responded, "You have to tell them what to say," provoking laughter from the audience.

This exchange prompted Rubbia to turn to me and say, "Those people, they don't care whether supersymmetry has been found, only the value of the masses." And after a short pause, to add, "Well, we succeeded in blowing their minds."

Gordon Kane followed with a decidedly nonpartisan talk, in which he said, "You'll see that I agree with a lot of what John Ellis says, particularly what he says about groups working besides mine and his. And that I don't agree with everything." He proceeded to flash another series of charts, with regions indicated for squark masses, which were allowed by the data, and other regions for gluino masses, not allowed by the data, and vice versa, according to the experimental data, about which nobody was sure at all. He showed how these data tended to uphold the merits of his model and discredit the others.

Finally de Rújula came on with his summary entitled "Should You Believe Any of the Above?" He gave an animated talk, pointing to members of the audience and waving his hands in the air. When he had finished, his message was clear: his was the only model that could explain the data. His final conclusion was that "News of the discovery of supersymmetry may have been fun, but is perhaps a wee bit premature."

The first serious blow to the UA1 monojets came in the discussion period that followed the SUSY talks. Until then, any speculation on their imminent demise had been based only on rumors and speculation. Then Guido Altarelli, one of the most eminent Italian quantum chromodynamics theorists, hit them with a beautifully staged and elegantly presented argument that proved to be the beginning of the end.

In some ways, it was not so much what Altarelli said, but how he said it. Unlike Glashow, who had jumped to his conclusions based on misconceptions about the state of the analysis, Altarelli wanted only to make a few points. But these few points turned out to be telling. They became known as the "Altarelli Cocktail."

Altarelli sat a third of the way back in the auditorium, and he turned so that he could address himself to the entire audience. He held the microphone in his left hand and waved his right hand in slow graceful circles to accent his words. He spoke distinctly, but in heavily accented English. He wanted to make it clear, he said, that it was not possible to explain what UA1 had seen in terms of a single standard model phenomenon. For instance, it would be easy to disprove that all the events came from Z's decaying into two neutrinos plus a gluon. However, he said, they could put together what he called a "wisely composed cocktail" of known things that *could* account for the UA1 signal. This cocktail might include two or three W's decaying into a tau and a neutrino, accompanied by a gluon, a couple of Z's plus gluons, the decay of a pair of quarks into two jets, of which the detector had mistakenly picked up only one, a particle disappearing into a crack in the detector . . .

The physicists laughed after Altarelli's mention of a crack in the apparatus. Rubbia liked to talk of his detector as hermetically sealed, without cracks. But this was not the case, and Altarelli had not pretended it was. Altogether, he said, this cocktail could account for all of the strange missing-energy events that UA1 had observed.

After taking care of last year's monojets, Altarelli then went after the dijets. The most impressive events in the data that Rubbia *had* shown were the ones in which there seemed to be two or three high-energy jets shooting out of the collisions. "So when you have two or three jets around," he asked, "how do you know that there is not some particle that goes in some crack of the apparatus? When you have a single jet, you can look around and say the jet is far away from the glue; it is far away from every blind region. But when you have many, there could be a particle almost anywhere. . . ."

In those high energy events he continued, it becomes more difficult to be sure that the missing energy is not simply the result of a particle that should have been detected, but wasn't, Rubbia had only five such events: "I counted them this morning in Carlo's transparencies," Altarelli said, "and five is a small number."

He handed the microphone back to the usher and turned slowly back around in his seat. He was finished. The Altarelli Cocktail seemed like an incontestable explanation for Rubbia's missing-energy events. Altarelli left the details of his cocktail to a pair of physicists, who would take care of them two days later.

WEDNESDAY

The Wednesday morning sessions covered the UA1 and UA2 upgrades and the experiments for future colliders. The afternoon session was on bread-and-butter physics from the collider experiments. It was the slow day of the week. Many of the physicists took the day off to go shopping in Milan or skiing at Cervino, on the slopes of the Matterhorn. They returned that night complaining that it had been wet, foggy, and even worse, flat.

Glashow drove down to Milan to do some shopping, and returned commenting that every time he passed a magazine stand, he saw Rubbia's face staring out from a prominently displayed copy of *L'Uomo Vogue*. He wondered why he hadn't made it to the cover of any fashion magazines when he won his Nobel.

Most of the UA1 gang returned to Geneva and to work.

THURSDAY

At three-thirty on Thursday afternoon Rubbia left the conference, on his way to Rome. An hour later, James Stirling and Steve Ellis took over the job that Altarelli had started.

Stirling was a dapper Englishman, in his mid-thirties, who had worked on QCD theory for years. Ellis was a bearded American of forty-one, who had come to CERN for one year "with the intention to get my hands dirty doing some phenomenology again . . . to work with the people who have the machine, because it's fun." When Ellis came to CERN, he and Stirling, together with a young Dutch theorist, started working on possible standard backgrounds to the UA1 missing-energy events. "I was just unhappy," Ellis said, "that I was hearing so much about supersymmetry and so little about what they were actually doing at UA1."

They had started working in January with the 1983 data from UA1 and UA2; both experiments had published events in which they had seen W's or Z's created accompanied by gluon jets. With these events, Ellis and Stirling had deduced what possible circumstances could occur to hide the W's and Z's when they were created, leaving only the jets visible, which would appear in the detectors as monojets. It was much like the work Denegri had been doing sporadically at UA1. They had worked through the laborious calculations of QCD that specified the frequency with which those circumstances would be expected to occur. They were doing nothing particularly original. They were simply doing more conclusively what other physicists had done in bits and pieces. By early February, they had results for the possible background to the monojets that were significantly larger than those UA1 had published. It was then that they decided they would discuss their work at St. Vincent.

Stirling spoke first, about possible effects from QCD—from the gluon and quark jets—in W and Z physics. He ignored the connection with the monojets. He had a soft British accent, and he worked calmly, letting the calculations speak for themselves, building up his case step by step. For every one of his theoretical deductions, he would check back with the experimental data to prove that the deduction had been valid. "Things seem to be in good shape," he would say, or, "This gives us encouragement." He was demonstrating that their theoretical calculations of the tail of the distribution accurately matched the available experimental data from UA1 and UA2.

"QCD," he concluded, "represents either a very serious, not very serious, or not serious at all background for monojets . . . and this is just a background to more exciting things."

Then Ellis took the stage. His lecture style was half comedian and half television sports announcer. He was funny, and often sarcastic: "I want to send a warning," he began, "that old-fashioned boring physics might account for the monojets. I realize that's radical and maybe unpatriotic in the place we're in, but let's do it anyway." And he proceeded with great zest.

The subtitle of his talk was "The Many Roads to Paradise." His first transparency read: "Remember the words of G. Altarelli, 'The sum of many small things is a big thing!' " And then he went through what he called his roads, one after another, which were simply the ingredients of the Altarelli Cocktail, now supported by actual calculations. He first calculated how many "monojets" might be expected from Z-zero's invisibly decaying into neutrino and antineutrino with a gluon jet thrown in. This was the one undeniable background to the missing-energy events. The UA1 physicists had calculated what to expect from this in their very first missing-energy paper. "There is no appreciable contribution predicted. . . ." the paper had said. Ellis, on the other hand, claimed that these could explain at least two of the UA1 events. ("This is actually conservative," he said. "It could be as great as four.")

He then proceeded with a series of calculations for the various "monojets" that could be expected from the ingredients of the Altarelli Cocktail. They were all small: a third of an event here, half an event there. But, as Ellis said, they were not zero. When he added them up—all the possible background contributions to what UA1 had reported as inexplicable by the standard model, all the events that could look like new physics even though they came from relatively mundane W and Z decays—they came to over six events. If he took into account the potential errors in his calculations, he added, that number could be as large as eleven, or as small as four.

"You may wonder whether that's a big number or not," he said. "It's a hard issue. It's certainly not zero."

By then, he had most of the physicists laughing. He was flipping his transparencies onto the projector and whipping them back off before the audience had time to read them. He had worked himself up into a frenzy, waving his arms in the air as though he were conducting an orchestra. He ended by saying that "I am not at all sanguine that a big fraction of that [monojet] signal is not just," his voice dropped dramatically to a near whisper, "plain old W and Z decay."

Much of the audience left without waiting for the last speaker. As far as UA2 was concerned, at least, that was it. Luigi DiLella turned to Pierre Darriulat and said, "It's too bad Carlo wasn't here to hear that." They seemed to feel that they had missed a good fight, because the irreverent tone with which Ellis had attacked the new physics might have enraged Rubbia.

DiLella turned back to me and said, "You might as well add to your book the lack of criticism that led to them being called new physics to begin with." And a younger member of UA2 said, "It was a dirty job, but somebody had to do it."

FRIDAY
In a summary talk entitled "Where Do We Go from Here?" Glashow, who a month earlier had appeared ready to resign himself to supersymmetry, now retrenched and gave what had to be considered his victory speech. The standard model would survive its latest threat. He had been around enough to know, at least, that whatever came out of UA1 this year, it would not be overpowering. Glashow gloated. He said that while they themselves had not come to any conclusion at the conference, the press already had. And he read from articles in *Science, Nature,* and "an even more distinguished journal," the *Harvard Crimson,* all of which claimed that the latest run ("That's the run that's kept secret from us," he said) had convinced physicists that the missing-energy events were, one, real, and two, probably proof of supersymmetry.

But supersymmetry, he said, was like syphilis; it was "the great imitator." He explained that the sufferers from supersymmetry, de Rújula, Ellis, and Kane, each had symptoms that contradicted those of the others. And he said that supersymmetry could never be proven false because "it is most precisely a theory that predicts nothing, and cannot explain anything." It could only be replaced by a superior theory. As new theories went, however, it was better than most. "The one theory that is clearly superior to supersymmetry," he said, "is the theory that is obtained from supersymmetry by deletion of all unobserved particles. This theory, a very exotic theory, is called the standard model. It has half as many parameters and half as many particles, and yet explains all of the phenomena of nature just as well.

"I think I have a conclusion here," he went on, speaking like a jury foreman presenting the verdict to the court. "I conclude, in view of the lack of absolute confirmation from the 1984 running of the existence of anomalous events, that there is no decisive evidence as yet that the standard model is either incomplete or incorrect. Indeed, the situation is more favorable to the standard model today than it was last year. . . ."

After Glashow's talk, the physicists broke for coffee. Out in the lobby, DiLella had a copy of the morning edition of *Corriere della Sera,* the Milan paper. He was translating an article to an audience of physicists, who stood around laughing and cheering him on. Hans Hoffmann, of UA1, Gordon Kane, and Steve Ellis were with him, and four or five others.

"Physicist Rubbia Now Wants a Second Nobel Prize," DiLella said, translating the headline. "The Italian scientist is trailing a strange phenomenon of missing energy . .

"Carlo Rubbia was asked by this guy here," DiLella explained, " 'What would you like to do next?' And he said, 'Get a second Nobel.' That's what he said. Huh? Quote, unquote."

DiLella stopped for a moment to let the laughter die down.

"Then it says that the Nobel Effect, which is generally quoted by scientists as being a fearful form of self-satisfaction, taking away any driving force or will, in Carlo Rubbia is working to the contrary and is trying to push him toward higher and higher enterprises.

"Then it says that Carlo Rubbia has spoken to the four hundred physicists who were listening to him like they were entranced.

"Two hundred twenty," DiLella said, correcting the article.

"But it always doubles," Hoffmann said.

"Like the number of Nobel prizes," added a third.

16.

MARCH, 1985:
ALL THE KING'S HORSES
AND ALL THE KING'S MEN

"Carlo is sort of the ultimate father-figure. In that sense, he defines what's right and wrong."
—Michael Levi, UA1 physicist

In the offices of UA1, it was as though St. Vincent had been nothing more than a bad dream. Maybe it had never even happened. On Friday night the usual gang were working late on the missing-energy analysis. Now Rubbia wanted the work done for a proton-antiproton workshop in Japan that would begin in ten days. Rubbia supposedly wanted to be able to show the Japanese some beautiful results and then sell them on the idea of pitching in with the next collider project at CERN.

Pauss and Honma were still optimistic about the monojets even after St. Vincent. They thought all the talk about the Altarelli Cocktail and the papers by Ellis and Stirling were simply irrelevant to the work they had in front of them. The theorists had had a year to suggest backgrounds to the experimentalists' new physics. Now the collaboration would work it out themselves and see who was right. "Those guys want to know if it's supersymmetry or not," Honma told me. "They want to make sure it's not obvious background. But their numbers are only coming from the theory. This isn't even necessarily close to the same thing that we actually measure. To me all that stuff is fine, but until you put it into the detector, it's meaningless; you can get any number you want. It won't bother me until we actually put the proper numbers and responses in. Then, if it comes out like they say, we have to worry."

The supersymmetrists were equally unperturbed by the St. Vincent outcome. The Saturday after the conference ended, I bumped into Dimitri Nanopoulos in the corridors outside the CERN library. Nanopoulos had not been in St. Vincent. He told me that he had heard through John Ellis that UA1 wanted to do the analysis right; they had no competition, they would take their time, and perhaps a final an-

232

swer would be out by the time of a conference on new particles that Cline had organized at Madison, Wisconsin, in early May.

Rohlf flew in from Boston that Saturday, complaining that he needed a break. He had taught both his own and Rubbia's classes the week of St. Vincent. He looked exhausted, with dark circles ringing his eyes. He didn't have much to say other than that he now acknowledged that they definitely had some backgrounds to deal with on the monojets—backgrounds they had not considered in depth until the theorists had brought them to their attention—but that he still felt that some of the events were simply too striking to be accounted for by the standard model. He was also untroubled by Ellis and Stirling's background estimate. "This guy's pulling numbers out of his ass," he said to me, as he paged through the transparencies from Ellis's talk.

Rubbia also arrived back on Saturday, screeching into the parking lot in his Mercedes. He called the missing-energy analysis group into his office and told them that they had one more bullet in their gun. They could either use it to prove they had a signal greater than the background, or they could turn it around and use it on themselves. As Rubbia saw it, the paper they would have to write should either definitively confirm that the missing-energy events could not be standard physics, or establish that they were compatible with the background. Nothing else would do.

Afternoon coffee in the cantina: Rubbia sat with Levi, Pauss, and Rohlf, as well as two ex-UA1 members who were passing through town, and an American physicist, who had worked on UA2 and was now working on one of the detectors—called CDF for Colliding Detector Facility—at the Fermilab Tevatron. Rubbia was baiting the American, telling him that every year since 1976, the physicists at Fermilab had claimed that they were three years away from getting beam, and now they suddenly say they are only two years away from getting it.* Rubbia construed this to mean that they were either finally making progress or simply getting frustrated.

Then Rubbia changed the subject. He mentioned the difficulties UA1 had had over the past year with the calibration of its equipment, and he said to the American, "You will need God on your side to do the calibration at CDF."

*In reality, Fermilab was colliding protons and antiprotons within six months.

"But, Carlo," the physicist replied, "you have God on your side."

"Maybe not lately," Rubbia said, "maybe we have the devil on our side. We have sold our souls, now we have to deliver."

Now that Rohlf was back in town, the gang was going to take one last collective shot at narrowing down the missing-energy signal. During February, Pauss and Honma had taken the five thousand events from the missing-energy trigger and honed that figure down to sixteen that seemed to have legitimate missing energy over 40 GeV. Anything over 40 GeV had a good chance of being new physics, something that could not be explained. Sixteen was a good number. The year before they had had five.

When Mike Levi returned from St. Vincent, they gave him the sixteen events that had survived and asked him to check them over carefully. Levi studied those events, and over the course of an evening he worked the sixteen down to three or four that he liked. As far as Levi was concerned, if simply by looking at the picture, he could propose a method by which the detector could have faked the missing energy, then the UA1 collaboration would never be able to convince other physicists that that was not what had in fact happened. Levi had a cynical sense of humor (among other things, he had a UA1 physics article rolled up on his bulletin board in a toilet paper dispenser), and neither Pauss nor Rohlf were particularly happy with the notes he had made. He had decimated their monojet and dijet sample.

On Saturday evening, Pauss, Rohlf, and Levi went to the Megatek to scan one by one the 1984 missing-energy samples. This was the first time that Rohlf had seen the final events. The first they looked at was Rohlf 's famous dijet of the previous October, the very first dijet they had taken. They spent almost an hour trying to measure the energy in one of the jets. The problem was that this jet had hit the inside of the detector at about 6 degrees from vertical, and the detector had openings at both the top and the bottom 5.5 degrees wide through which the cables and wires for the CD were drawn out of the detector. Rohlf said that if the jet was electromagnetic energy, all of it would be detected by the calorimeter cells along the crack, and they could be sure that nothing had snuck through. Levi disagreed, and argued to toss it away. He couldn't believe that they were even debating the point at all. The jet was real close to the vertical, and the large missing-energy arrow was pointing away diametrically opposite it. They'd never be able to prove to anyone that an energetic particle hadn't shot right out that crack.

After an hour of rotating the image on the screen and adding masses and looking for tracks in the CD, they decided to pass it on to the Saclay people, who would double-check the energy in the calorimeters, and who, they suspected, would probably kill it instantly.

They moved on to the next event. It had two jets, back to back, one of which had tracks that seemed to be going directly into and through the cracks between the cells of the hadron calorimeters. You could see them clearly on the Megatek. The missing energy was opposite the jet, and Levi again argued that the odds were good that the missing energy was caused when something snuck right down those cracks. This event, Levi said, was one of the few that he at first liked, but he didn't anymore. Rohlf said it certainly could be no more than a 15 or 20 GeV neutrino causing the missing energy, which was not so exciting.

At this point, they decided to flip quickly through the remaining events to get an idea of what they had in store, and then go eat dinner. The next event appeared on the screen: a huge three-jet event with 58 GeV of missing energy.

"There's nothing wrong with this event," said Rohlf. "I know this event well."

Levi laughed. "Would you take a look at the calorimeters," he told Rohlf. "There's a hadronic shower of 61 GeV and no tracks leading to it. Half the detector was off."

He pointed to one of the slices that made up the hadron calorimeter. This was the part of the detector that would register impacts by any of the particles, called hadrons, that contained quarks, and sure enough one null directly opposite the missing-energy arrow registered 61 GeV. There were no tracks in the central detector anywhere near the cell, which there would have been if that huge energy deposit had been left by a charged particle. Either it had been some kind of neutral particle that did not register in the central detector, or that single cell had somehow misfired.

"Could have been a neutron," Rohlf said.

"Great," Levi replied, sarcastically. "We've discovered neutrons. I thought Chadwick already did that in 1932."

Levi and Rohlf were good friends, and Levi loved the bantering, particularly when he had the advantage, if Rohlf hadn't had time to study the events and he had. Now Levi wandered off to talk with Mohammadi, leaving Rohlf and Pauss to quietly flip through the events. "This is a W to electron," Pauss said of the next event, and, "This is a tau," of the following one. And so on.

Levi wandered back and the three decided to get dinner. It was ten o'clock. As they put their coats on, Rohlf wrapped it up. "Looks like we're left with about four or five good events," he said. "We expected twice that many, eight or ten, and we got half. That's compatible with statistics."

"It's also more compatible with background," Levi added. "After you throw out all the events that you know how they're bad, you're left with the four or five that you don't know how they're bad."

After dinner, Rohlf and Pauss decided to give up for the night. Levi went back to the Megatek. He called up the three-jet event with the huge missing energy and the huge reading in one calorimeter cell. The neighboring cells registered a few GeV of energy each—the closest even had 16.5—but the hot one had 61 GeV and no tracks. "So what's the problem?" said Levi. "It's got to be a real winner to explain this." He studied it for a while, looking at the CD tracks, looking at the event from different angles. And then he gave up. "I don't know," he said quietly. "I don't know."

Afterward, Levi told me that if Rubbia was going to be at the upcoming missing-energy meeting, he would have to get somebody else to present his results. There would be no way that Levi could do it with any conviction. Levi had once before "implied that the UA1 detector was not a divine temple as believed, or as sold," and he had paid for it.

A few weeks earlier I had watched him give a status report on the events that the computer had apparently reconstructed incorrectly. One category of such events he had labeled "Disasters." Rubbia, who had walked out before his talk and walked back in midway through it, wanted to know what a disaster was. When Levi explained that these were events with apparently incurable problems, Rubbia blew up.

"So why call them disasters?" he shouted. "I don't like that word. I don't want you to use it. I don't like anybody who takes this detector as a joke. Don't tell me this is laughable. It's not. If something is wrong, we want to know what it is." Two minutes later, when Rubbia asked Levi how a Z candidate could be missing one of the electron tracks, Levi's irreverence betrayed him again. "Because it's a disaster," he said.

"If you use that word again, you leave this room," Rubbia replied. "You leave this group."

One of the problems that Levi had when he arrived at UA1 was dealing with authority, which meant Rubbia. Levi had grown up in the academic environs of the University of Chicago, where his father was first provost and then dean of the university. He had done his undergraduate work at Harvard (where he worked part time at Rubbia's Brookhaven experiment, and where Rubbia had once temporarily fired him for answering back during one of Rubbia's rages.) He then went on to get his doctorate at SLAC on the Harvard contingent of the Mark II experiment. That collaboration, run by Burt Richter with Gerson Goldhaber and George Trilling of Berkeley, had the reputation of doing some of the best physics and turning out some of the best young physicists.

Levi had arrived at CERN a year before. He had started at UA1 in March, just as they were pushing to get the new physics paper out. "You're like anybody else coming in from the outside," he said. "I didn't even know how to use the Megatek, let alone be critical. You learn to be critical."

Now he was critical, which was one of the best characteristics a young physicist could have. "I had the role of being the cynic," he told me later. "That's an easier role to be in, because instead of creating this thing, you get to punch the holes in it." But Levi had grown up in the atmosphere of American physics. He didn't see why, if he had something worth saying, his voice shouldn't be heard. Levi thought himself a good physicist, and he was. But he had trouble adjusting to the fact that he needed diplomacy more than anything else in order to make his talents known at UA1. As Sadoulet explained, "In Italy, a professor is somobody you do not contradiut too easily. In Germany also. But in America, yes, I can picture a young graduate student telling Dick Feynman [a Nobel Laureate] that what he says is crap."

Although Levi found the atmosphere at UA1 stifling compared to that at SLAC, he also thought it to be in some ways ideal. He felt like a character from a version of the old kids' tale who, after skipping the sowing, the reaping, the milling, and the baking, gets to eat the bread with the little red hen, provided he keeps his mouth shut while he does it. "Obviously, I could never put together that kind of money and that kind of experiment and do that kind of physics myself," he told me. "Carlo's a vehicle. It's a phenomenal situation. You come into an experiment, and these people are sitting around scratching their heads and they don't quite know what to do, and you can jump into this confusion and try to piece it back together. That's what you want to do. That's the great pleasure of doing physics."

*   *   *

Pauss and Rohlf were back on the Megatek again at five o'clock on Sunday afternoon. They both looked grim. They had spent much of the afternoon talking with Rubbia. Pauss had a cold. Rohlf still looked like jet lag had the best of him. Levi was just back from skiing and joined them late. Levi worked the keyboard, still wearing his ski pants. They ran through the same events, slowly checking tracks and fits, calculating jet energies and track energies and masses, looking at photomultiplier counts, and looking closer and closer at CD hits, trying to refit tracks.

They stared at one event with a massive jet going into one side of the detector. The cells in the hadron calorimeters were all blindingly bright, denoting huge deposits of energy, and on the opposite side the cells were also lit. But in front of the calorimeter cells, in the gondolas, there was nothing at all. Levi said it looked like somebody had forgotten to turn on half the gondolas. On top of that, the missing energy was exactly horizontal, which meant that a particle could easily have slipped off through the central detector without leaving a track and then sliced through the cracks between the calorimeter cells. If it had done so, this would explain the missing energy.

"I somehow recall," Levi said, smiling, grandstanding a little, "when I was young and innocent on this experiment, sitting in awe of the great Jim Rohlf, and an event like this came up, and you spent about five seconds on it and said it's an obvious kill."

After a few minutes of looking at it, Rohlf suggested that Levi might be right.

"Let's go to the next one," said Levi.

"The problem I have with this event," he said, "is there's 600 GeV of energy seen, and 65 not seen. So there's 665 GeV in this event." That was enough to make a handful of W's. It could happen, if they had actually had two collisions in the same event, two protons colliding with two antiprotons. It was not wholly unlikely. Rohlf asked if that was what could have happened. But Levi said no, there was no sign of the two vertices that would mark two collisions.

"But this is one of the best events we have . . ." Rohlf said.

They finished scanning at one-thirty in the morning. Levi said that they had two or three monojets with their gold-plating still intact when they were done. Maybe one dijet. Rohlf said they had four to five monojets. They both told me that this meant nothing until they had calculated the background. And Rubbia had told them that they had to have the background done by the time he left for Japan on Friday.

Monday morning Rohlf walked over to the CERN publications office to collect a handful of promotional leaflets. He wanted to show them to his kids back at Harvard, and Rubbia had also asked him to get some for the physicists and politicians in Japan. On the way back, Rohlf went to the cantina for coffee and bumped into Keith Ellis, a theorist and an old friend from Fermilab. Ellis was in town like many others after St. Vincent, hanging around for a few days before flying back to the States.

The two sat down at one of the oval tables in the middle of the cantina. Ellis, a Scotsman, had studied originally in Rome under Altarelli, and had worked in both Europe and the States. For the last three years he had been doing QCD theory at Fermilab.

They began to talk of old friends, and then Rohlf asked Ellis a question that he needed answered to help him calculate the background for the monojets. It required a good knowledge of QCD theory. One of the local theorists had told him that Ellis would know it off the top of his head. Ellis didn't, and they batted it around for a few minutes until finally Ellis suggested asking Stirling. That set the conversation off on a different track, since Stirling worked with Steve Ellis, and they were the ones who had been throwing background at UA1 in St. Vincent.

For the next twenty minutes, Keith Ellis cross-examined Rohlf. All Ellis wanted to know was what every physicist wanted to know: What was going on at UA1? Rohlf insisted that they had nothing to hide. Both Ellis and Rohlf worked on the same wavelength. They didn't hurry to answer.

"Those guys don't know anything," Rohlf said, about Steve Ellis and Stirling.

"That's because you guys don't tell them anything, Jim," Keith Ellis replied.

A long pause.

"So is it junk, what they're saying?" Ellis asked.

"Yeah," said Rohlf. "We have some beautiful events. We just don't know what the background is."

"So they're trying to help you."

Pause.

"So when are you going to tell us?" Ellis continued.

"Listen," Rohlf said. "This is a difficult experiment. Other people take a year to analyze the data. We should take two or three by their standards."

"The thing I don't understand about what was presented at St. Vincent," Ellis said, "was why Rubbia said everything was confirmed. It was clear that when he wrote 'Confirmed,' he meant that it was still new physics. I think there's some duplicity there. If you don't understand the background, then you can't say it's confirmed."

There was another pause. Ellis leaned back. He rested his arms over the two chairs on either side of him. Rohlf leaned forward, playing with a now empty coffee cup.

"I have to go back and report to the boys at Fermilab," said Ellis. "The DOE is paying a lot of money to send me to this cafeteria. What do I say: 'Events confirmed'?"

"You better take back Carlo's transparencies," Rohlf told him.

"You're full of bullshit, Rohlf. You really are." He said it one friend to another.

A long pause.

"Okay," said Ellis. "Very good."

A longer pause.

"We'd like to do some more work," said Rohlf.

"I'm not going to stop you. Go right ahead."

"Let me ask you one general question of the meeting," said Rohlf. "Did you get the impression that the missing energy was placed in an uncertain state?"

Ellis nodded.

"Why?"

Ellis said nothing.

"Did you think the [Anomalies] scorecard was dirty?" Rohlf asked.

"Of course I did."

"People said that the scorecard was below the belt," Rohlf said.

Ellis shrugged.

"That's the way life is," said Rohlf. "No one can fault us for not saying anything."

"Why was the stuff talked about in the United States before Europe?" Ellis asked.

"That's not true," Rohlf replied. "The first presentation anywhere of the data was in Santa Fe, where Carlo showed a few of the events. But the second presentation was at Rutherford. And that's what started all the bullshit."

"It's all over the Italian papers," said Ellis. "As well as *Gazetta dello Sport* [an Italian soccer paper]. Carlo's in there getting a medal."

Pause.

"So the question is, Jim," said Ellis, "when are you getting into *L'Uomo Vogue?*"

Long Pause.

"Yep," said Rohlf. "We should have done something. But Carlo gave a talk, which I understand was extremely short. People are just a little too ignorant."

"People are too ignorant?" Ellis was a little amazed.

"About what was happening. You have a right to get conservative. You get these spectacular events and people get nervous and back up to stage zero. Pretty soon we'll be at the state we were at last fall. That's my prediction."

Pause.

"So are you going to become spokesman of UA1 when Carlo retires?" Ellis asked, smiling.

"I wouldn't touch it with a barge pole."

"Sounds like a good job," said Ellis. "Lots of travel. Freedom of action."

"I think I'll take a leave of absence next year . . ."

On Monday afternoon Rohlf and Pauss scanned the lower-energy monojet events, those that had between 35 and 40 GeV missing energy. They hoped they would find a few more potentially inexplicable events, but they found nothing other than a couple of tau candidates.

When I talked to Rohlf later that day, he was obviously feeling a bit beleaguered. My presence didn't seem to make him any happier. Rohlf was nervous about both the missing-energy events and my book. He joked about how long it would take the world to forget them both. When he was back in Boston, Rohlf told me, he had asked de Rújula how the missing energy would rank by comparison with the high-y anomaly, if it turned out to be wrong. De Rújula had said they would be neck and neck.

I kept thinking of something Levi had said to me a few days before. "Could you imagine," he said, "working on something for a month and then telling Carlo you went up a blind alley and it didn't work? Never."

Rohlf was now at the point where if he took a skeptical position on the physics, the way Levi did, he could endanger his position on the analysis. Whereas Levi could afford to be skeptical, Rohlf couldn't. Or not yet. When I asked him about the scanning he had done the night before with Levi, he said, "I'm just careful. Levi's just a kid; he just joined the experiment."

One leading U.S. physicist later explained the situation to me in a more general sense. "Carlo is a terrible tyrant," he said. "In his tyrannical mode he will not accept criticism from people in his own group. This, I think, is the cause of his impetuous physics results: nobody dare criticize him when the analysis is going on. You can't say, 'Carlo, you forgot this, you're wrong.' "

What the physicist didn't say was how this affected the young people working under Rubbia. The very first time I had talked to Rohlf, back in October, he told me that his ex-colleagues at Cornell had warned him not to work for Rubbia. Everything will be going fine, they told him, and then you'll make one mistake, and Rubbia will never remember any of the good things you did. "It reminds me of that Steinbeck line about Salinas Valley," Rohlf had said. "In Salinas Valley, when it's hot, it's hot as hell. You can never remember that it was ever cold. And when it was cold, you can never remember that it was ever hot."

Rohlf was not the only one to feel that pressure in Rubbia's experiment. In January, Daryl Dibitonto had quit the collaboration and gone off into industry. He had gotten his Ph.D. at Harvard under Rubbia, and had been a disciple in some sense for ten years. On his last day at CERN, Dibitonto had been bitter. Standing in the hallway outside his office, he had told me that he remembered Rubbia saying once, "Physicists are like lemons, you squeeze them for all they're worth, and then you throw them away."

Now Rohlf was trapped by Rubbia's philosophy of physics. He had no choice. He knew that if they were to do the monojet analysis thoroughly, he'd have to keep on Rubbia's good side, and this meant being optimistic, at least in front of Rubbia.

Rubbia wanted to just push the physics out. Damn the background, and full speed ahead. He believed that you couldn't linger over an analysis or the physics community would lose interest. Even if you were right, you'd lose the impact. So Rohlf was pushing the analysis as hard as he could. He knew that they had committed themselves last year, and before he admitted any pessimism, he would make sure the analysis was done thoroughly. As Cline had said back in November, Rohlf "gave his talk at Leipzig on the new physics and his ass is on the line." But it was more than that. If the missing energy really was new physics, it was the most important discovery they could ever make.

When Martin Perl had discovered the tau in the mid-seventies, he had had trouble even though the tau was created in one out of every five collisions of electrons and positrons. It was an enormous signal compared to that of the monojets, which were created in one out of every billion collisions of protons and antiprotons. "Parts per billion physics," Perl himself had told me, "is a new era, and people have to learn. They may be able to do it, but it may be beyond our limited brain power. If you don't try, you don't know."

Now Rohlf was saying that they were not in bad shape. They had a small signal. They would calculate the background, and whatever happened, they would live with it. "I bet you we come out looking real clean with this missing-energy stuff," he told me.

On Tuesday afternoon the wet snow of the past few days turned to a gray drizzle. Everything and everyone was quiet, subdued either by the weather, the workload, exhaustion, or all three. The previous day Rubbia had told Pauss that they had better have the background done by Wednesday, when he left for Harvard where he would stop before going on to Japan. Pauss passed the word along.

The physicists waited in the conference room for Rubbia. Rohlf had gone back to Boston that morning. Pauss was sick; Levi was trying to decide whose cold he was coming down with; Mohammadi also had a cold. Michel Della-Negra, an Annecy physicist who had done much of the work on the top quark analysis, was there. Levi was leaning back with his arms crossed and yawning. The others were scribbling on transparencies. The meeting was scheduled for two-thirty. At three Rubbia walked in, saying, "Sorry, folks."

Mohammadi began the meeting with his Monte Carlo for the W to tau decays. He hoped to estimate with the Monte Carlo how many W to tau decays they would expect over 40 GeV in their monojet sample. He had only the 1983 data. The 1984 data, he said, would be done by tomorrow. He showed his plots and Rubbia made his assessment.

"So a cut at 40 GeV," he said, "is leaving essentially nothing from tau decays."

"We should wait until we see the 1984 data," Mohammadi cautioned.

Rubbia leaned back, his hands behind his head, his feet up on the desk.

"If you have one tau decay above 40 GeV," he said, "your hands will be cut off."

"Then I resign right now," Mohammadi said, smiling.

"If you resign now, something else will be cut off."

Richard Batley, the Englishman from the trigger crew, came next with what they called event mixing. Batley, an amiable young blond from Queen Mary College in London, was duplicating in practice what Ellis and Stirling had done in theory. Batley had created a computer program that would take parts of real W's—in the form of hundreds of thousands of words of computer code—and randomly combine the various parts to see if they could fake monojets. The program took the electrons and the jets and the missing energy in the W events and spun them around within the detector as though they were on Roulette wheels. In doing so, it could create thousands of hypothetical W's, instead of the two hundred the experiment had taken, of which a reckonable percentage would have an electron buried in the jet and invisible—leaving what the physicists would call a monojet.

Batley's program included all the theoretical predictions as well as all the vagaries of the detector. At least, all that they were aware of. He flashed a transparency of what he had done so far, demonstrating that his program worked. He said he'd like to check it some more, and then run it on the 1983 and 1984 data. Pauss asked him if he could do so by tomorrow.

"I can do it," he said. "I can go through the motions. But I won't have great confidence in it."

A young German physicist reported next. She had been conscripted just days earlier to calculate the expected background from the invisible Z decays. She also had not yet produced any concrete results. Rubbia ran through the simple calculations to show her roughly what she should expect and why it should be easy. It was one of his effortless displays of dexterity, performed like a magician doing a few simple sleights of hand just to keep his fingers supple.

Z's should decay into two electrons only 3 percent of the time, Rubbia explained, and into neutrinos 18 percent of the time. So there should be six times as many neutrino decays, which are invisible, as electron decays, which are visible. Now, he said, they had approximately twenty Z decays into electrons. Six times 20 is 120, so they should have about 120 invisible Z decays into neutrinos. Now for the next step: According to the standard model, Z's and W's must act in the same way. Since they had two hundred W's, of which only one came with a jet over 40 GeV, they should expect only 0.6 Z's into neutrinos with a jet over 40 GeV. And that would be their background—it would be insignificant.

It was all very simple the way Rubbia had laid it out.

"So," the young German said, "we don't have to worry about it?"

And Rubbia slapped his hand against the desk, suddenly yelling at her.

"No, goddamn it, why do you say that?" Then his voice dropped, like a disappointed father. "You don't know how to play poker."

Pauss spoke last, giving the results of the scanning of their sixteen missing-energy events. She talked directly to Rubbia, looking a little disconsolate. She first showed a chart listing the various characteristics of all the events, then went through them one by one, flashing pictures of the events as she went.

"One is the famous monojet found in this run," she explained. "It's really a very super-nice, clean event. The second one also looks extremely clean . . . there are at least four tracks in this event with energy above 1 GeV. In the next event, you can at least reconstruct three tracks without any problem and at least four soft tracks. The mass is compatible with the tau. . . . The next event has one stiff track. It's compatible with tau decay . . ."

Rubbia interrupted here, "These events are really beautiful," he said. "They're really absolutely gold-plated and hard to explain."

Pauss continued. The next was a W, then a final monojet with problems that they just couldn't understand. And then she hit the two-jet events. The first was a gold-plated 65 GeV dijet. The second was the dijet that Rohlf and Levi had argued over, with the jet too near the vertical to be safe. The rest, she said, she would skip because it would be boring.

Rubbia said, "Okay, c'est finis," and walked out.

The next day Rubbia went to lunch with Pauss to discuss the monojet situation. Later, Pauss complained that she had to get Rubbia what he needed for his talk in Japan and at the same time prepare herself for a talk she had scheduled in Munich. She would leave for Munich Tuesday morning, attend an official dinner that night with local politicians, and then present the talk on Thursday. She would have to be back at CERN by Sunday night. Rubbia, she said, wanted to say more on the missing energy in Japan than he had in St. Vincent, but he was relying on her to get it for him. The Wednesday deadline was out. Now they were aiming for Friday night. Rubbia would leave for Japan that day from Boston, but would call CERN at midnight Geneva time, just before he left.

When I asked Pauss if she thought they could still convince the world that the monojets were new physics, she said, "I think so. I mean, this is what we will try. We're doing a much more elaborate background calculation than last year, because now we have all this criticism, and we know what to do about it."

She believed, as did Rubbia, that perhaps the previous year they had had a statistical fluctuation of these mysterious monojet events. That, for example, they should have had only one or two last year, but they were lucky and got five. This year, when they figured they should get fifteen good ones, they had only three or four. I asked her about the criticisms that had come from physicists within the group. It seemed at times that she and Rohlf and Rubbia were the only ones who weren't ready to admit they'd made a mistake on the monojets. Some of the collaboration had told me that after the paper had come out the previous year, they had looked at the five monojets and decided that only two of them were not obviously taus or junk, which would contradict the argument about statistical fluctuations.

"If somebody has really strong doubts," Pauss told me, "why don't they say, 'Listen, I don't believe that.' Why don't they sit down and make a calculation and prove that all this stuff we're doing is junk or garbage or whatever? They don't do this. They just say that they don't believe it. I don't care if people believe. What we're trying to do is a good job on the background. We know how to do that. This is what we're doing. . . . There have always been people who are so scared and so disbelieving in anything that they would cut everything down."

On Thursday, the work continued. Pauss and Honma were joined late in the afternoon by Charling Tao, who had taken the TGV in from Paris. Since the blowup in February, the Parisians had been integrated into the analysis, and Tao was doing a Monte Carlo similar to Mohammadi's. The two stayed in touch by phone and computer link-ups.

When I saw Tao and Pauss late in the evening, they were in mild hysterics from lack of sleep and long nights and were laughing uncontrollably. I asked them how it was going, and Pauss said they had some beautiful events. Tao said, "Yeah, beautiful taus, and beautiful Z's decaying to neutrinos, with a gluon thrown in." And they laughed. I asked if they were as confused as I was about this, and Pauss said no, they definitely had some nice events. Tao said, "No, they definitely had beautiful tau candidates." They laughed again.

Rubbia's call from Boston was expected at midnight on Friday, but by early afternoon they had nothing final. They could still Telex him last-minute results once he arrived in Japan, however. Mohammadi was going to stay until his Monte Carlo was finished, and he would have background estimates from W to tau decays, W to electron and Z to neutrinos. He might have them by one in the morning, he said, or he might not have them until five. But he would stay to see.

The results, at least the preliminary ones, were inauspicious. Richard Batley's event mixing seemed to point to one or two events where the electron was lost in a jet. This roughly matched what Ellis and Stirling had calculated. The young German physicist's results pointed to one to three events over 40 GeV of background from the invisible Z decays. She complained that nobody wanted to listen to her, that nobody liked her results because they were too high, and that they wanted her to redo them. She also said there were a lot of uncertainties. Charling Tao's W to tau calculation showed one to two events over 40 GeV.

By now, it was obvious that whatever happened, the background was not nearly as trivial as they had made it out to be in the previous year's monojet paper. Levi told me that when Rubbia had hurried that paper to get in the same issue of *Physics Letters* as UA2, he had done so on the strength of a W to tau Monte Carlo that had been three weeks old. "You come into a situation where you have something you have to explain and you're just getting your Monte Carlo going," Levi said, "then you're not in a situation where you can be making sophisticated calculations. There's really something wrong. . . . If they had had a really good W to tau Monte Carlo, they would already have been worrying last year, just on those preliminary numbers." In addition, the possibility of a background source to the monojets from W's hidden in jets—which was what Batley was working on—had not been calculated the previous year, or if it had, had been calculated quickly, providing a negligible answer.

Mohammadi, who had the definitive background Monte Carlo, was still not finished. "I don't believe anyone's numbers," he told

me. "It's all preliminary. The only numbers I believe are my own."
And Levi added, "Nobody else believes your numbers. They only
believe their own."

I passed Hans Hoffmann standing on the balcony outside his
office with Rudy Bock, both of them senior physicists at UA1. Hoff-
man had heard the preliminary numbers on the analysis, and he said
that even figuring that nobody had any definitive word yet, there
were a maximum of only four good monojets, and that was not good.
Bock said the signal had become compatible with the background.
Hoffmann added, "It is still compatible with supersymmetry, but
that and five cents, as the man said, will get you a cup of coffee."

Later in the afternoon, I gave Tao a ride to the Geneva train station
to catch the TGV back to Paris. Along the way she talked about the
rules of science, the so-called scientific method, about how it dictated that if some experimental result appears to be contrary to the
accepted dogma—in this case the standard model—they must first
assume that somehow that result had been misconstrued, and explore every avenue by which that could have happened until they
were finally forced by necessity to accept its validity. But, she said,
that was simply not how it worked in practice.

"Every one of us," she said, "if we're stupid enough or foolish
enough to work on this, is really looking for something new. Every
one of us would like it. Even I am disappointed that I don't see a
clear signal from monojets. Not because I have any illusions about
the events of last year. But you always think it would be nice to find
something which is not predicted, which is not expected.

"I'm talking about what you wish is there," she went on. "If you
were working that hard and crazy like this, it's because you wish
there's something, that what you're doing is not for nothing. Now,
for example, if what we're finding this day is confirmed, that means
that all the data can just be explained by trivial background. All the
numbers are heading this way. That's physics. That's life. I feel
disappointed, of course, but I will have to say, this is it. That's what
we measure; that's we've done."

At midnight, Pauss was in her office waiting for Rubbia's call and
to tell him that they had nothing final yet. Mohammadi was in the
computer room squinting at the computer screens, trying to will the

computer to go faster. Nobody else was in the building but the cleaning woman.

At five-thirty in the morning Pauss went home. Mohammadi finished a half hour later. Rubbia never called. The final numbers hadn't changed. The background and the signal were equivalent.

It was beautiful that day in Geneva. "I might go skiing tomorrow," Mohammadi told me.

EPILOGUE: DESERT REDUX

"At the moment I do not feel I have the right to write another wrong paper based upon these experiments, which I no longer believe."
—Sheldon Glashow, March 1, 1985

In Japan, Rubbia said nothing publicly on the subject of monojets. He returned to CERN in mid-March, threatening to fire the physicists responsible for the analysis. They stayed out of his way until his temper cooled. He then set them to reworking the analysis. The deadline this time was the Washington meeting of the American Physical Society in late April, and if not by then, the conference on new particles in Madison, Wisconsin, in early May.

In the third week of April, the executive committee of UA1 strongly requested that Rubbia reconsider his decision to resign as spokesman of the experiment. Rubbia agreed to stay on.

One week later, in Washington, Rubbia claimed that perhaps half of the monojets were consistent with W to tau decays; the remainder still appeared to be inexplicable. In Madison, he forbade his physicists to bring transparencies to the conference or to the UA1 group meeting that preceded it. He also forbade them to discuss the state of the analysis with anyone, even the UA1 collaboration members who had not directly participated in it.

Honma, Pauss, Levi, Mohammadi, and Batley had been working for two months without a break. By May, the signal and the background were still equivalent. "It was the Altarelli Cocktail," Levi told me later. And he added about their analysis, "It was definitive. It was the state of the art of what we could do. It was not Megatek work. It was serious honest to God work."

In the process, they had created a program that could, with good probability, identify W to tau decays. They had run through it the events from the 1984 "new physics" paper. Those events, according to this refined analysis, were compatible with W to tau background.

They still had no answer that satisfied Rubbia. The work continued.

The deadline then became the International Europhysics Conference in Bari, Italy, in July. One of the young physicists told me, "Carlo will have us working on this until we beat it to death or fuck up and there's a signal." Shortly before the Bari conference, Rubbia requested a complete presentation of the analysis. Rohlf, who had rejoined the analysis after the school year had ended, reported on the group's work and was castigated while doing so by Rubbia. Rubbia told Rohlf to concentrate on the electronic readout system for the microvertex detector, which was the responsibility of the Harvard contingent and was in danger of not making the deadline for the fall installation.

Rubbia said nothing about the missing energy in Bari. (He said nothing about the top quark, either, the existence of which had also become dubious at best after the analysis of the 1984 data. On the way to Bari, Rubbia told Luigi DiLella that they would have no top quark presentation because the wife of the physicist who had prepared the talk had complained that her husband would be away from home too long, and he had cancelled his presentation at the last minute. When I later mentioned this to the uxorious physicist concerned, he smiled and responded, "There were complications, but not the ones you heard.")

Alvaro de Rújula, who presented the summary speech for the theory section, came up with a cartoon of a man going to a job interview:

"Name?" the interviewer asks.

"July 23, 1945," he responds.

"Profession?"

"10 Main St., Meyrin."

"Any congenital defects?"

"Pi r squared."

"Perfect," says the interviewer. "You can start work for us right now!"

In the last frame of the cartoon, the man is installed behind a desk and a sign reading: "Information on UA1 monojets."

At the end of July, Rubbia had the monojet analysis presented to the collaboration. It was called a physics seminar, and took the form of a status report on unfinished work.

To his working group, Rubbia now seemed to be conceding that the monojets had disappeared. The background, after three more months of analysis, was slightly greater than the signal. But he told those working on the analysis that they would not rush to write the paper. They now had the equivalent of a copyright on the word, he told them, and should monojets reappear next year, they did not want to have a paper attributed to them that stated they were gone. Rubbia did not want another alternating neutral-currents situation. And he told the physicists to consider the political implications of the paper. Several of the younger physicists interpreted this to mean that if the premature announcements of the discoveries of monojets and the top quark had indeed had much to do with the success so far of Rubbia's future projects—the LHC and his proton-decay experiment in Gran Sasso—then any public disavowal of their existence could endanger the viability of those projects.

The key word was "public." The physics community had no delusions on the analysis. The word had spread, through one channel or another, that the UA1 monojets were dead. Finally, in Kyoto, Japan, at an international symposium in mid-August, Rubbia made the analysis public.* In a one-hour presentation with 100-plus transparencies, Rubbia spent no more than ten minutes each on the top quark and the missing-energy analysis. He covered the work so quickly that few in the audience could follow it. On the monojets, he concluded that "There is a clean signal from tau events, but there may be something additional, a question which can only be resolved with more statistics."

His claim was not convincing, however. Both Roy Schwitters of Harvard and John Ellis of CERN concluded in their summations, according to the CERN Courier, that "the monojet signal from the UA1 experiment . . . can be accounted for without too much effort."

Later, when I asked Schwitters what, in his opinion, Rubbia had in mind by insisting he might still have a signal, he replied, "It's just Carlo, is what it is."

When the storm had cleared, the desert was back in all its parched magnitude. Supersymmetry, once again, had not a single speck of evidence to tie it to the mundane specifics of the real world. But in the meantime, the course of theoretical physics had taken another

---

*A week earlier, Rohlf had given a somewhat more realistic talk at an APS meeting in Eugene, Oregon. Pauss was scheduled to present the analysis at CERN on October 1, but Rubbia later canceled her talk.

turn. The hottest candidate for a potential ultimate theory of every-thing was no longer supersymmetry or supergravity, but something called superstrings.

Two theoretical physicists, John Schwarz of Caltech and Mi-chael Green of Queen Mary College in London, had put together a theory that postulated that the fundamental particles of the universe were not pointlike, as electrons and quarks supposedly were, but were infinitesimal strings, joined at their edges into loops, that could vibrate and rotate in different modes to make other particles. Super-string theory had for some time been nothing more than a curiosity. It could be made to work only in mathematical equivalents of ten dimensional universes. But in the summer of 1984, through some mathematical miracles, it had been made to work, at least in ten dimensions. No other theory that tried to link gravity to quantum physics ever had. It also appeared to naturally include supersymme-try, hence the name superstrings.

By early 1985, the theorists seemed on the verge of constructing a superstrings model that would work in four dimensions. If they could do it, they believed they would most likely come out with a supersymmetry theory not too dissimilar from the scenarios that had been batted around CERN for the past few years.

If superstrings was correct, then supersymmetry was correct. Even more extraordinary, if supersymmetry could be proven, then physicists would have the ultimate theory of everything, and the long search would be over. Dimitri Nanopoulos told me that when he heard about the progress that had been made, he couldn't sleep because of the excitement. "We don't care anymore about UA1 results," he told me, "because we know we will have it."

By the summer of 1985, supersymmetry, courtesy of superstrings, was established more solidly in the dogma of theory than ever. All that was needed was some shred of experimental confirmation. But that wouldn't be coming from UA1.

On August 4, 1985, I sat in the cantina at CERN drinking beer with Alvaro de Rújula. We talked about whether the demise of the monojets had created a corresponding lull in the supersymmetry work. Or whether the theorists were so hot on superstrings that they would continue on supersymmetry undeterred as well. De Rújula predicted that 90 percent of the theorists would work on super-strings and the connection with supersymmetry, because it was fashionable. When he intimated that this was not a healthy state,

I asked him what he would prefer to work on. Rather than answer directly, he digressed.

"It must be remembered," de Rújula told me, "that the two people most responsible for the development of superstrings, that is to say, Green and Schwarz, have spent ten to fifteen years systematically working on something that was not fashionable. In fact, they were ridiculed by people for their stubborn adherence to it. So when people come and attempt to convince you that one must work on the most fashionable subject, it is pertinent to remember that the great steps are always made by those who don't work on the most fashionable subject."

"The question then," I said, "is what do you work on instead? What will your next paper be on?"

"That's a question for each theorist to ask himself," he replied. "And it depends on whether you want to survive as a theorist, or you have the guts to think that pride in your own work is more important than the momentary recognition of your fashionable contribution. That's for each person to decide by himself, depending on his level of confidence in his own genius."

"So," I repeated, "what is your next paper going to be on?"

"I'm trying to tell you," de Rújula said, "that I have no idea."

GLOSSARY

Antimatter the counterpart of matter; a particle with identical mass and spin as a particle of matter, but with other properties, such as electric charge, reversed: i.e., the positively charged positron is the antimatter counterpart of the electron.

Baryon a particle composed of three quarks, such as a proton or a neutron.

Boson one of two basic species of particles defined by a characteristic called spin. Bosons have only integral values (0, 1, 2 . . .) of spin, and include photons, W's and Z's, pions and gluons.

CDF Colliding Detector Facility, an experiment at the Fermilab Tevatron accelerator.

CERN the French acronym for the European Center for Nuclear Research, located near Geneva, now called the European Laboratory for Particle Physics.

Charm one of the six types, or "flavors," of quarks.

Collider an accelerator that drives two separate beams of particles into head-on collisions.

Color an attribute of quarks and gluons analogous to electric charge.

Cosmic rays energetic charged particles, predominantly protons, that continuously bombard the earth from space.

Decay the transmutation of an unstable particle into one or more lighter particles.

DESY Deutsches Electronen-Synchrotron Laboratory, the electron accelerator complex in Hamburg, West Germany.

Dijet a collision between subatomic particles in an accelerator in which two jets of secondary particles are detected shooting from the collision in one direction, and nothing in the other.

Electron an elementary particle that carries the fundamental unit of negative electric charge.

Electron volt a unit of energy equal to the energy acquired by an electron in falling from the negative pole to the positive pole of a one-volt battery. Also a measure of mass: the mass of a proton, for example, is about a billion electron volts (GeV).

Electroweak theory   the theory that encompasses the combined action of the electromagnetic force and weak force.

Elementary particle   a particle that is not a composite of other particles.

Event   an interaction between subatomic particles.

Fermilab   Fermi National Accelerator Laboratory, a proton accelerator complex west of Chicago.

Fermion   particles with spin equal to half integers, such as protons, neutrons, quarks and electrons.

Flavor   generic term for the various species of quarks, of which there are thought to be six: up, down, strange, charm, bottom, and top.

Gamma rays   high-energy photons.

GeV   one billion electron volts.

Gluon   a massless particle that carries the strong force.

Graviton   a hypothetical massless particle that carries the gravitational force.

Hadron   a particle composed of quarks and affected by the strong force.

Higgs boson   a hypothetical particle required in the electroweak theory to give mass to W and Z particles.

Integrated luminosity   the total number of collisions generated between the particles in the two beams of a collider, measured in inverse nanobarns.

Intermediate vector bosons   the particles that carry the weak force, denoted by W+, W— and Z^0.

ISR   the Intersecting Storage Ring, a proton collider at CERN.

Jet   a collimated stream of secondary particles shooting from the high-energy collision of two particles in an accelerator, usually resulting from the interaction of quarks or gluons.

J/Psi   a particle consisting of a charm quark and an anticharm quark bound together.

LEP   the Large Electron Positron accelerator under construction at CERN, scheduled for completion in 1988.

Leptons   a species of particles that, unlike quarks, do not interact through the strong force. They include the electron, muon, tau, electron neutrino, muon neutrino, tau neutrino, and their antiparticles.

Linear accelerator   a device that accelerates particles in a straight line.

Luminosity   measure of the intensity of particles in the beam of an accelerator.

Meson   a particle consisting of a quark and an antiquark bound together.

MeV   one million electron volts.

Monojet   a head-on collision of subatomic particles at the CERN collider, in which a single jet of secondary particles is detected shooting from the collision on one side, and nothing is detected on the other.

Muon   a heavier cousin of the electron.

Neutral currents manifestations of the weak force in which particles interact through the exchange of the Z^0.

Neutrino a neutral, apparently massless elementary particle, which interacts rarely with other particles.

Neutron the neutral constituent of the atomic nucleus.

PEP Positron Electron Project, the electron-positron collider at SLAC.

PETRA the German acronym for the Positron-Electron Tandom Ring Facility, an electron-positron collider at DESY, near Hamburg.

Photon the massless, chargeless particle of light and electromagnetism.

Pion a meson whose exchange between protons and neutrons binds the nuclei of atoms together.

PPBAR Proton-Antiproton colliding beam device. (Also known as **pbarp**.)

Proton the positively charged constituent of the atomic nucleus, composed of three quarks.

Quantum Chromodynamics (QCD) the theory describing the strong interaction between quarks and gluons

Quantum Electrodynamics (QED) the theory of electromagnetic interactions.

Quarks the elementary particles that compose protons and neutrons and all other hadrons.

Resonance an extremely short-lived state—too ephemeral to be a particle —formed in high-energy collisions between particles.

SLAC the Stanford Linear Accelerator Center, an accelerator lab near Stanford University in California.

SLC the Stanford Linear Collider, an accelerator scheduled for completion in 1987, which will smash together electrons and positrons

SPS Super Proton Synchrotron, a proton accelerator at CERN.

Standard model the overall picture of high energy physics representing a consensus on particles, forces, and theories

Storage ring a ring-shaped device that circulates beams of particles in stable orbits for hours or days.

Strong interaction the strongest physical force, which binds quarks together into neutrons and protons and those into nuclei of atoms.

Supergravity an unproven theory that unifies all forces within a single mathematical model.

Superstring theory a new and unproven theory that describes elementary particles as being flexible, one-dimensional strings.

Supersymmetry an unproven theory based on a symmetrical relationship between fermions and bosons that postulates a universe of yet-unfound particles.

Tau a heavy cousin of the muon and electron.

TeV one trillion electron volts.

Tevatron   an accelerator at Fermilab that produces collisions of protons and antiprotons at a total energy of 2 TeV—inaugurated in late 1985.

UA1   Underground Area One, both the large hall and the experiment within the CERN antiproton-proton collider.

UA2   Underground Area Two.

Upsilon   a particle consisting of a bottom quark and an antibottom bound together.

W particles   positively or negatively charged carriers of the weak force, along with the Z^0.

Weak force   the fundamental physical force responsible for certain types of radioactive decay and other particle decays.

Z particle   neutral carrier of the weak force, along with the W's.

ACKNOWLEDGMENTS

Enormous numbers of people are needed to build, maintain, and run the machines and experiments of high-energy physics. No book could encompass them all, or even come close. For this reason, I want to offer my thanks to the hundreds of graduate students, technicians, and engineers of the collider project and the experiments who, by definition, do all the work and yet are not mentioned in the book. And I want to offer my apologies and thanks to those physicists and engineers who are mentioned only briefly, but not at all in proportion to their contributions, which were simply incalculable.

I would like to thank the management of CERN in general, for allowing me to live at the laboratory for six months as though I belonged, and Harald Bungarten, Roger Antoinne, Gwendoline Kordu and Mirella Keller in particular, for six months of patiently enduring my interminable questions and requests. I would like to thank all those members of the physics community—again too numerous to mention—who gave me their time and consideration and offered advice and counsel, especially those who read various portions of the manuscript and did everything in their power to make sure I would be as accurate as they could make me.

I would like to thank Derek Johns, for his incomparable editing and guidance, and Ann Adelman, for her editorial assistance. My special thanks to Bruce Schechter for his knowledge, suggestions and help all along the way. To Kristine Dahl for many "mushes" and for being the best agent a young writer could have. And most of all, I would like to thank my brother for lending me the benefit of his intelligence when I ran short of my own, my parents for their infinite patience and support, and Marion, for just about everything.

ABOUT THE AUTHOR

GARY A. TAUBES was born April 30, 1956, in Rochester, New York. He studied physics at Harvard University, aeronautical and astronautical engineering at Stanford, and journalism at Columbia. He is currently a contributing editor to *DISCOVER* and has written on boxing for the *Atlantic* and *Playboy*.